Praise for *Foraging as a Way of Life*

There's always something new to learn about wild plants and their uses, and Mikaela Cannon brings a combination of personal experience and dedicated research to her engaging profiles of edible species. With detailed descriptions and photos, she brings to life food and craft uses; fun recipes for syrups, jams, and lozenges; and relevant lore about common, weedy plants, as well as some native species special to western North America.

— Jared Rosenbaum, CERP, author, *Wild Plant Culture: A Guide to Restoring Edible and Medicinal Native Plant Communities*

Reading *Foraging as a Way of Life* is like walking through the forest with a good friend. Mikaela Cannon has created an incredibly engaging, expansive, and insightful book full of personal experiences to guide you through a full calendar year of walking in the wild. If you have one field guide, one cookbook, one mushroom manual, and one foraging calendar, make it this one!

— Terryn Corbett, herbalist, and owner, Om Naturale

What sets this foraging guide apart from others is that it is arranged by months and seasons to direct you to the best plants and mushrooms to forage at any time of the year! I appreciate the clearly indicated caution boxes and "similar species/look-alikes" sections provided for each entry, and the many practical uses, recipes, and products to make from your foraged finds. This will be a fun book to plan my next foraging trip with.

— Jeanine Davis, associate professor of horticulture, North Carolina State University; author, *Growing and Marketing Ginseng, Goldenseal and other Woodland Medicinals*; and co-owner, Our Tiny Farm.

Foraging for wild foods is the most ancient human activity, the original way of life. Mikaela Cannon's passion for harvesting nature's abundance has yielded a comprehensive guide to foraging throughout the year. Beautiful photographs, detailed profiles, and numerous recipes make this book an exciting guide to reclaiming a naturally healthful diet.

— Darrell E. Frey, Three Sisters Farm, and author, *Bioshelter Market Garden* and *The Food Forest Handbook*

Congratulations for creating this invaluable resource. Mikaela Cannon's observations and knowledge of the great outdoors provides the foundation of this guide. Her *joie de vivre* is reflected with photos and an in-depth narrative that is designed to be shared with foragers and would-be foragers of all ages.

— Cam Luckock, *Your Country News*

In *Foraging as a Way of Life*, Mikaela Cannon has succeeded in weaving together the most important features of a practical and easy-to-use field guide with the crucial heart-informed elements of a text which is cherished, re-read, and shared throughout one's life. Our world needs more works like this which inform and inspire us to live more aligned with nature. With each season, each harvest, and each preparation we can make choices that either affirm or deny our connection to all life on this planet. This book shows us how.

— Sarah Elliman, author, *A Tree Just Like Me*

A plethora of ancient knowledge permeates these pages, complimented by Mikaela Cannon's deep connection to the land, and her expertise in plant medicine. Her passion and wisdom help the reader to deeply resonate with what has always been true—that powerful medicine is available from nature.

— Tracee Andrews, author, *Whispers on the Ocean*

Foraging truly *is* a way of life for Mikaela Cannon, and this book captures her meticulously researched knowledge, practical experimentation, and deep respect for the earth and its Indigenous knowledge-keepers. It is beautifully illustrated, and includes recipes, medicinal uses, personal stories, detailed plant descriptions, and wise precautions. Though organized seasonally for her part of the world, many of the plants and mushrooms highlighted can be found throughout North America. Mikaela's advice is good medicine for a world suffering from the effects of industrial agriculture and disconnection from nature; take even a small dose and make your world better.

— Remy Rodden, biologist, award-winning envirosinger,
and environmental educator www.remyrodden.com

Foraging as a Way of Life

foraging [as a way of life]

a year-round **field guide** to wild plants

MIKAELA CANNON

Foreword by Nancy J. Turner

new society PUBLISHERS

Cover design by Diane McIntosh.
Cover images: © iStock (#1024367670, 1166553806, 1179077064, 1481357287). Back cover photos: Indra McMorran.
All photos, unless otherwise specified, by Mikaela Cannon.

Printed in Canada. Second printing November 2024.

Inquiries regarding requests to reprint all or part of *Foraging as a Way of Life* should be addressed to New Society Publishers at the address below.
To order directly from the publishers, please call 250-247-9737 or order online at www.newsociety.com.

Any other inquiries can be directed by mail to:

New Society Publishers
P.O. Box 189, Gabriola Island, BC V0R 1X0, Canada
(250) 247-9737

LIBRARY AND ARCHIVES CANADA CATALOGUING IN PUBLICATION

Title: Foraging as a way of life : a year-round field guide to wild plants / Mikaela Cannon ; foreword by Nancy J. Turner.

Names: Cannon, Mikaela, author. | Turner, Nancy J., 1947– writer of foreword.

Description: Includes bibliographical references and index.

Identifiers: Canadiana (print) 20230587631 | Canadiana (ebook) 2023058764X | ISBN 9780865719972 (softcover) | ISBN 9781550927900 (PDF) | ISBN 9781771423861 (EPUB)

Subjects: LCSH: Wild plants, Edible. | LCSH: Wild plants, Edible—Identification. | LCSH: Cooking (Wild foods)

Classification: LCC QK98.5.C3 C36 2024 | DDC 581.6/320971—dc23

New Society Publishers' mission is to publish books that contribute in fundamental ways to building an ecologically sustainable and just society, and to do so with the least possible impact on the environment, in a manner that models this vision.

Certified B Corporation

MIX
Paper | Supporting responsible forestry
FSC www.fsc.org FSC® C016245

Contents

FALL

WINTER

Disclaimer

This book is intended to be educational and informative. The author and publisher disclaim all responsibility for any liability, loss, or risk that may be associated with the application of any of the contents of this book. Although the author and publisher have made every effort to ensure that the information in this book was correct at press time, the author and publisher assume no responsibility for any errors, inaccuracies, omissions, or other inconsistencies within this book and disclaim any liability to any party for any loss, damage, or harm resulting from such errors or omissions.

Territorial Acknowledgment

As a recent immigrant to this area, I stand in awe of the Secwépemc and the Syilx-Okanagan Peoples. They have lived here for over 10,000 years and have an extraordinary knowledge and connection with this land. They have been excellent land stewards for an immense amount of time. I recognize their knowledge and wisdom as deep and profound, and I am grateful to learn from them.

Their land and human rights were stolen by the colonial powers, and it is time to right these wrongs. I support social justice for Indigenous Peoples through self-determination and decolonization. I strive to learn of the history, speak up against injustice, and I pledge to support the First Nations communities around me to the best of my ability. I vow to carefully examine my thoughts and actions in order to recognize when I am acting entitled or from a place of privilege, and to correct my actions toward greater understanding and respect. It is my deepest wish that my actions be of benefit to the land and to the people.

Foreword

by Nancy J. Turner

"Foraging" might suggest "random picking and plucking," but effective foraging is far from that. There is so much to know about the species you are harvesting and their environments: Where do they grow? Which are edible, which are poisonous, and what parts are usable? What other species are similar and might be mistaken for them. What is their harvesting season? How can they be harvested sustainably, without reducing their populations or impacting other species? How are they prepared? What insects and animals are they associated with? Are there cultural protocols around their harvesting and use?

Mikaela Cannon provides all this information for the plants and mushrooms she forages, based on meticulous research and—most importantly—her own experiences. For her, the book reflects a compellingly personal journey, filled with the joy of "*finding nourishment from the land*." Her enthusiasm is infectious; you want to participate in her world: "*In my opinion, there is nothing as delicious as chanterelles fried in butter and eaten on toast. Before cooking them, I gently clean dirt away.*" You want to taste, and make for yourself, the unique dishes she has created and provided recipes for: *Mallow Lasagna*; *Mountain Ash Turkish Delight*; *Seed Bar with dock seeds, hazelnut catkins, and nettle seeds*; and *Spruce Tip Gummies.*

Foraging as a Way of Life, written with deep appreciation and acknowledgment of the Indigenous Peoples within whose lands and waters Mikaela resides and thrives, highlights their priceless knowledge and the immense gifts of knowledge keepers like the late Secwepemc Elder and ethnobotanist Dr. Mary Thomas. Writing with humility, and recognizing the wrongs of colonialism, and the environmental destruction caused widely through agriculture, clearcut logging, and other recent human activities, Mikaela still conveys a sense of wonder, fun, and adventure, of gratitude and love for the natural world. She includes key information about insect pollinators and animals that depend on these plants, with warnings to always think of their needs, not just our own: "*Remember that this plant* [saskatoon] *is a favorite of the bears too. Make sure that there is plenty left for them.*" She also inspires us to want to learn about new species, their ecological requirements, and their cultural history, and not to be afraid to experiment with new recipes. Building relationships with the plants she is harvesting and using, she shows a deep caring, an essential "kindness" and kinship towards the species featured. Her own words say it all: "*When reciprocity is your guiding beacon, the forest becomes your community, and you feel supported by it. Your neighbors (the plants, animals, and water) become friends.*" Thank you, Mikaela, for sharing your foraging adventures with all of us!

— Nancy J. Turner, PhD, CM, OBC, FRSC, FLS;
Distinguished Professor Emeritus,
School of Environmental Studies,
University of Victoria

How to Use This Book

The top heading shows you the month. I feel like the beginning of the year ought to be March, so I have started the year of our book with that month. Each season has three months, with five plants and/or mushrooms every month. Each season has its own color: spring, green; summer, violet; autumn, orange; and winter, blue.

Title

The name of the plant or mushroom, its Latin name, and the family it belongs to will be in the title.

Synonyms

Other common names for the plant or mushroom are listed here.

Description

This section describes the features of the plant or mushroom so that you can be comfortable in identifying it. **Some parts are bolded** to make it easy to find the information you need at a glance.

Uses

Here I describe how our family uses the plant for food. Included are recipes. Take my measurements with a grain of salt...

Other Uses

This part highlights additional uses, such as medicine and material made from the plants and mushrooms. Here you will find ecological information and other interesting facts.

Similar Species / Look-alikes

To give you confidence in your identification, this section explains how to tell species apart.

Caution

If there are special concerns about the plant, they are listed here. Do not skip this part!

Photos, Drawings, and Captions

The photos, drawings, and caption information are meant to highlight specific features of the plants and mushrooms to give you confidence in identification. Photos and illustrations are by the author unless otherwise credited.

Glossary

If a word or term is unfamiliar to you, search for it in the glossary. This section also includes drawings to further aid understanding.

References

Here you can find the literary and online resources I used while writing this book.

Index of Names

The plants are listed in alphabetical order under both their common name and their Latin name. The synonyms are not listed.

Meaurements

Most measurements in this book are expressed in metric, using centimeters (cm). One inch is approximately 2.5 centimeters. If you would find a conversion tool helpful, there are many good ones online, such as thecalculatorsite.com.

Welcome to This Book

First, I would like to say how excited I am that you have picked this book to read. I hope with all my heart that this book will be of service to you.

Who I Am

My name is Mikaela Cannon, and I am grateful to live and forage on the traditional and unceded territory of the Secwépemc and Syilx-Okanagan. I live in Spallumcheen, BC, in western Canada, and I forage in both the Shuswap region and the Okanagan Valley. I am a recent immigrant to this country, and I vow to learn how to live respectfully here from the Peoples whose ancestry on this land goes back at least 10,000 years. I encourage you to learn about the Indigenous Peoples and culture of the land that you are on. How can you find ways to understand, support, and respect Indigenous self-determination and uplift the voices of that Nation? Below are some links to gain knowledge: native-land.ca. This site has both Canadian and global information; nctr.ca/records/reports. National Centre for Truth and Reconciliation. Information on truth and reconciliation reports and other resources; firstvoices.com. Learn about Indigenous language and oral culture.

I grew up in Sweden with a Canadian mother and a Swedish dad. I remember fondly hanging out in the woods making forts and eating wild plants that I had gathered. I do not know how I knew that they were edible; no one had told me. Since then, I have become much more careful and systematic in how I use wild plants, see my methodology below.

Most of my early adult life I spent travelling, learning about the local plants of each new area. Together with my partner, Peter, we finally settled down in the Okanagan, and this place is where my heart is. The Okanagan is an abundantly rich region with many different biotopes and with great biodiversity.

Even if you do not live in the Okanagan, you will find that many of the plants covered in this book will be found in your area too. I have deliberately chosen plants that are widespread and some that are considered invasive. I have seen many of these in the United States, Europe, Scandinavia, South America, South Africa, and Australia.

For more than a decade, I have been foraging and teaching people about plants in my region. Everything I do boils down to one thing: to protect and care for the nature around us. After we had children, it became natural for them to accompany me on my forest excursions, and to take part in preparing food and medicine from wild plants, as well as learning how we can live sustainably on the land. I set about writing about the foraging we do as a family, with the hope that our children, when they are older, will have a record of the knowledge gleaned, and a glimpse into how they spent their childhood. That was my initial premise. When other people found out about my project, they said that they wanted a copy when I was done. The wonderful folk that came out to my courses kept asking me when this book would be made available. This made me realize that there is a real yearning to connect with nature.

My Methodology

I want to give you, dear reader, an insight into my methodology so that you can judge for yourself if you should trust what I have to say. When learning about a new plant, I consult as many varied sources as possible. I like to find a fair number of scientific papers, and I double check the references. I also consult various internet sites that I trust, again checking their sources. When I am able to, I take workshops and classes, as well as speak with Elders, who are fountains of wisdom. I read any and every field guide and book on plants that I can get my hands on.

For me, one of the most important parts of learning about a plant is spending time with that plant, really getting to know it. To be able to identify the plant properly, I make detailed notes for myself with drawings. If I can, I study the plant in situ for a few years to be able to recognize it in all its stages. I note to myself who eats it and who it grows next to. I take photographs and catalogue them for later reference. After I have gone through these steps and decide to eat a plant, I record how I prepare it and how I feel after eating it. Over the years, as I come across new information about the plant, I add it to my folder. Often my folder includes a pressed specimen too. When I use a plant medicinally, I record how effective it is in order to remember what worked and what did not.

My hope is that this book can help you find the confidence to identify the plants and mushrooms here described, and thus be able to find nourishment from the land. The book is designed so that each month covers five plants or mushrooms, to give you an idea of when we harvest them. Some can be harvested in different seasons for different parts, and I have included that in each chapter. The book can be read in any order, or from cover to cover. Each individual chapter will explain how we use the plant or mushroom. See section, How to Use This Book, page xv, for a breakdown of the format of the book.

I am not an herbalist, a nutritionist, a botanist, nor a doctor, so I cannot tell you how you should use any of the plants for food or medicine. I can, however, explain how my family and I use them. In the book, I have tried to be as clear as possible in the description of the plants, how we use them, and potential look-alikes, but I am in one location, and you dear reader, might be in another location with different look-alikes, ones which I am unaware of. If there is one thing I know for certain about plants and their uses, it is that I do not know everything. Please be careful if you are picking plants for food and medicine. If in doubt—leave it out. Always make sure you have correctly identified the plant or mushroom you plan to use. Ensure that you are in a location where foraging is allowed and welcomed and be cautious to not harvest from polluted areas. Do not take any unnecessary risks!

Sustainable and Ethical Harvesting

Another important aspect of foraging that cannot be stressed enough is that we need to harvest sustainably, or not at all. My rule of thumb is to never pick the very first specimen that I find. Rather, at that first plant, I pause and take a moment to look around. Is this plant abundant? Do the plants look healthy and robust? Is this species vulnerable, red-listed, or common? If I determine that it can handle harvesting, I gently take a small portion. Never take every plant. Make sure to leave some to reproduce into future generations and serve as food for animals. If you are uncertain of how much is too much, take less. If it looks like you never touched it when you leave, then you are on the right path.

I also ask permission of the plants to harvest them. I state my intention to the plant, and then I wait for the response. It takes becoming still and aware. Initially, I was unsure of what a yes felt like and what a no felt like. I felt a little silly, to be honest. But once I felt a clear no, I did not question it anymore. For me, a no feels like a huge tiredness and heaviness in my body. A yes, on the other hand, feels like a lightness and giddiness. Here is where some readers might be skeptical and think that I am too airy-fairy in my approach. That is ok. Despite what you think about talking with plants, taking a moment to slow down and contemplate that you are about to take a "resource" (to use Western scientific-sounding terminology) should give you pause. Is this the best time to harvest this plant? Do I have the energy to process what I harvest so that it does not go to waste? How much can I actually use? How much do I need? These are all important questions to ask before embarking on a harvesting spree. Taking the time to slow down and ask yourself these questions can minimize damage that could otherwise be done from actions before thought. In this, I speak from experience. I have in the past been so excited to harvest a great abundance that I skipped this step of checking in, only to later realize that I harvested more than I needed, and some ended up going to waste. Intentionality can help prevent this.

For many thousands of years, Indigenous Nations have thrived through intimate connections with the web of life, tending and cultivating the wild plants of their land. Before you forage, respectfully consider whose land you are on, ask permission where necessary, and if possible, learn the Indigenous place names and plant names where you are harvesting. Many Nations are experiencing lack of access to their traditional herbal medicines, and some of their important plants are being overharvested by unsustainable practices of careless foragers. Make sure that your harvesting is never detrimental to place, people, plants, and animals.

Finally, I give an offering to the plant/plant community for its gifts. We are truly indebted to plants on this planet. Without them, there would be no life. We literally owe our lives to them! I acknowledge this as I make my offering. An offering can be anything. I sometimes bring water, sometimes dried herbs, at other times I might offer a song. When I can, I pick up garbage and dispose of it safely. Always, I offer to be a spokesperson for habitats at risk. An amazing thing happens when you interact with the forest in this way. When reciprocity is your guiding beacon, the forest becomes your community, and you feel supported by it. Your neighbors (the plants, animals, and water) become friends. It is no longer a green chaos; now you can spot individual friends that help you when you are sick or hungry. You see abundant nurturing. You belong here. This planet is your home.

With Kindness,
Mikaela Cannon

Notes About the Land

The plants in this book can be found in plant hardiness zones 3 to 8 (from parts of Canada to the southern US states of Georgia, New Mexico, and California. The areas I have visited for this book are partly on the unceded and traditional lands of the Secwépemc (Armstrong, Enderby, and the Shuswap) and partly on the unceded territory of the Syilx Okanagan People (Vernon and Kelowna). I have spent more time on the plants in the Secwépemc area which is why a greater focus is spent on their stories and history. Both the Syilx and the Secwépemc have been living in the Okanagan for over 10,000 years. I am grateful to both Nations for their knowledge and kindness.

Over thousands of years, they came to know the land intimately, named the formations they saw, and created a flourishing culture based on respect and a spiritual relationship with all living things. Marianne Ignace and Ronald E. Ignace explain beautifully in their book *Secwépemc People, Land and Laws* how they use a sustainable resource harvesting system, which is supported by a moral system that recognises all beings as important. Humans are not considered masters over other species but rather responsible beings in a mutually beneficial web of life. Both the Secwépemc and the Syilx have extensive knowledge about which plants can be used as food, medicine, and material. This knowledge is passed down from generation to generation by knowledge-keepers. The people actively manage their environment by replanting roots of plants, pruning bushes, weeding, and breaking up the soil, all to ensure a more abundant harvest. The late Mary Thomas, a renowned Secwépemc knowledge-keeper, referred to the Secwépemc plant management and harvesting "just like a garden."

The governance of the settler society that came to BC in the late nineteenth century declared that the Secwépemc's use of the land was primitive. They could not see that the Secwépemc cultivated the land. This shows how biased the colonial powers were to their own ideas and systems, rather than open to learning a valuable and sustainable way to make a landscape productive and healthy. The Secwépemc and Syilx People moved through the landscape in seasonal rounds and carefully maintained the plants and habitats they came across. A seasonal round was made up of five distinct seasons: early spring (snow melting), mid-to-late spring (root gathering), summer (berry and high-elevation root and medicinal plant gathering), late summer to early fall (salmon season), and mid-to-late fall (hunting season). Any harvesting that took place was done with great respect for everything given. Nothing was wasted, and prayers and giving thanks accompanied harvesting occasions. Thanking the creator, or animal, or plant for giving you food shows your humble dependence on the natural world in order to survive.

The gold rush on the Fraser and Thompson River in the late 1850s brought people from afar to search for riches. With them, they brought a sense of entitlement and a smallpox epidemic that killed two-thirds of the Secwépemc population. Settlers spread over the land and were granted titles to land by a governing body that had not asked permission from the very people who had lived in the area for over 10,000 years. When British Columbia joined the Canadian Confederation in the 1870s, the Indian Act was put in place and further restricted Secwépemc and Syilx Peoples from receiving rights to their land and robbed them of civil rights. Find out more in the book *21 Things You May Not Know About the Indian Act: Helping Canadians Make Reconciliation with Indigenous Peoples a Reality* by Bob Joseph.

Since the onset of settler colonialism, many plants and animals have come close to extinction. Cattle grazing has destroyed many root crops, depleting former harvesting sites. Suppression of the use of fire to manage the landscape has caused its own slew of difficulties, and lost access to land meant that people were deprived access to their traditional medicine and food. The inherently genocidal practice of stealing children from their families and forcing them into residential schools created an immense loss of connection to the Elders and community, and much language and important knowledge was lost because of this intergenerational trauma. There has been one atrocity after another inflicted on the First Nations People of this area, but they are still standing. They are still strong. They remain wise. Indigenous People all over the world are teaching Indigenous knowledge and culture, and are encouraging settler decolonization. I encourage you, dear reader, to do whatever you can to support them in their initiatives. Learn about the people whose land you forage on. Respect them. Raise their voices. Listen, advocate, and take action.

Crystal D. Morris is Splatsín te Secwépemc and Tsartlip of Saanichton Nations. She is the niece of renowned knowledge-keeper and ethnobotanist Mary Thomas. Crystal has followed in Mary's and her grandmother's footsteps, taking on the role of knowledge-keeper of plant knowledge. Crystal gathers traditional herbal medicine and creates teas, salves, and other products, which she sells out of her store, Sharing Mela'hma. Her extensive knowledge about plants and traditional medicine, combined with her understanding of cultural practices and cultural safety, allows her to provide wholistic care. Prior to starting her own business, Crystal ran the Splatsín Health Centre, and she holds a diploma in business administration, a certificate in aboriginal health care management, and more recently, a diploma in herbalism. Both Crystal and her mom Adele (Ethel) Thomas work tirelessly to bring awareness and change to our communities. Adele, a residential school survivor, shares her stories and wisdom to bring about truth and reconciliation. Her words are powerful and important to heed. Adele's hope is to create a safe space for the survivors' truth to be heard and for unfair policies, which still suppress Indigenous Peoples, to be terminated. It is crucial that truth and reconciliation takes centre stage, and that healing can come to the Peoples of this land. Both Adele and Crystal speak of their deep rootedness to the land and the strength of their People.

Land was stolen from Indigenous Peoples all over Turtle Island, and their way of life was disrupted. It is now time to truly learn from them how to live right on this land; in reality, the well-being and survival of us all might depend on it. It has been documented that land managed or co-managed by Indigenous Peoples have the highest biodiversity of all lands, including parks and nature reserves. It is time to right all the wrongs that a colonial hierarchy and Eurocentric worldview has brought to First Nations societies. It is up to all of us to do our part.

As an immigrant to Canada, I vow to learn as much as I can from the Secwépemc and Syilx culture. I vow to speak up about injustice when I see it. I vow to show respect and take the time to learn about the culture and the language from the people of this land. I am willing to take an honest look at myself and see how I might subconsciously act in an entitled or racist way and strive to correct my actions and speech toward greater respect, deeper understanding, and genuine compassion. I seek to take part in Indigenous initiatives that will bring about systemic change, and I aim to be guided by love in all that I do.

In 2021, 215 suspected unmarked graves of children were found at the former Indian Residential School in Kamloops, and it sparked searches at other residential schools for remains of children who never came home. To this date, 4,400 suspected unmarked graves have been found in Canada. Many of the people in the Splatsin community have attended residential schools, some never came home, and the community is feeling the trauma.

To help commemorate the residential school survivors and the children who never came home, the community is creating a monument in their honor. If you want to support this project, the Splatsin Tsm7aksaltn Society (Splatsin Teaching Centre) is looking for donations and support. A portion of my earnings from this book will be donated to the Splatsin Tsm7aksaltn Society. If you feel like contributing and want more information, see splatsin.org/monument.

I hope that the information in this section is accurate, and if I have made mistakes, they are mine and mine alone. Should I have made errors, I would love feedback on how to correct them.

A Year of Plants and Mushrooms

Wild Ginger

Asarum caudatum | Aristolochiaceae

Synonyms

Long-tailed ginger, western wild ginger

Description

Ground-covering plant in moist habitats. Grows from a **rhizome**, at each node, there are two dark evergreen **leaves**. They are slightly hairy, heart shaped, and long-stalked. **Flower** grows at the end of a rhizome and is often hidden under the leaves. Flower is hairy, dark maroon, and has three long sepals that each taper to a point. Each sepal is generally longer than 3 cm. **Seeds** are dispersed by ants who are attracted to an oil on the outside of the seed. The entire plant emits a strong scent when crushed. Pollinated by flies.

Asarum means "hazelwort," and the genus was given this name by Linnaeus. *Caudatum* means "with a tail" and refers to the long sepals.

Figure 5.1. Wild ginger plant.

Uses

It is exciting to see wild ginger poking out from under the newly melted snow, the leaves from last year still green. The leaves are not as fresh and exciting to use at this time, but the rhizome is a lovely spring treat—potent and fragrant. We are fortunate to have an abundance of wild ginger here on the farm, but it is in danger of disappearing entirely in many places; never over-harvest this or any other plant. We usually only eat small amounts of it. You can use the root/rhizome fresh or dried; I prefer the more flavorful and less soapy taste of the dried rhizome. A little portion of it added to a stew imparts wild and wonderful undertones. I use it as a spice rather than as a vegetable. It can also be brewed into a tea.

The western North American wild ginger (*Asarum caudatum*) has edible leaves that have a similar taste to the rhizome, so they can replace it. This is better for the plant, as it can recuperate from the loss of a leaf easier than a root piece. I often add a few leaves and stalks to soup, stew, or stock. These can also be used in place of the roots to brew a pleasant tea. The flavor is like commercial ginger (*Zingiber officinale*) but with more wild nuances.

Other uses

Wild ginger plants are reported to contain aristolochic acid, which is toxic to the kidneys. For it to do damage, you need large amounts for an extended period. Avoid using this plant in vinegar or in alcohol as more aristolochic acid

Figure 5.2. Note the long tapering sepals.

Figure 5.3. Rhizome.

Figure 5.4. Top of leaf.

is then extracted. Traditionally, wild ginger was cooked or extracted in water only. If you are worried about ingesting too much aristolochic acid, using the water the plant has been steeped in and discarding the plant matter might minimize the amount ingested. Wild ginger is lauded for its antibiotic properties and helps with stomach troubles and indigestion. The leaves are antifungal and antibacterial. I use it as a tea to promote sweating when I want to flush a cold out of my system.

Caution

Do not use if pregnant. Do not consume in large doses over extended periods; see text above on aristolochic acid. Be extra cautious if you have impaired kidneys. People with sensitive skin may develop rashes from the leaves. Please harvest this plant sustainably as it is in danger of disappearing in many places.

Similar species / look-alikes

- **Eastern wild ginger** (*Asarum canadense*) Note that this is a different species than western wild ginger (*Asarum caudatum*). Eastern wild ginger is reported by some to have poisonous leaves, while others claim that they are edible. The root is edible in both species. I do not have any experience using *A. canadense*, so I cannot comment on its edibility. Leaves are deciduous, and the sepals of the flower are shorter, less than 2.5 cm long.
- **Early blue violet** (*Viola adunca*) The best way to tell the two apart is by the flowers. The blue-purple violet flowers stick up above the plant and have five petals. The smell of violet differentiates it from ginger. Edible. See page 17.

Weeping Willow

Salix babylonica | Salicaceae

Synonyms

Babylonian willow, Peking willow

Description

Native to northern China, common as an ornamental landscaping tree (up to 20 m). Grows near water. **Bark** is thick, furrowed, and grey. **Trunk** is relatively short, and its gracefully drooping branches often touch the ground. **Branches** have yellow/green bark, the **buds** look like little yellow duck beaks, flattened against the branches. **Leaves** are narrow and pointy (5 cm or longer). They are dark green and leathery on top, light-colored underneath, with finely serrated edges. Early in spring, yellow **flowers** are borne on catkins. The tree is dioecious; female and male flowers are on separate trees. Pollinated by bees and other insects and is an important early spring food for the bees.

Salix means "to spring or leap." *Babylonica* means "of Babylon." Babylon city is today known as Hillah and is in Iraq.

Figure 6.1. Willow under the weeping willow.

Figure 6.2. Catkins in spring.

Uses

The young shoots, flowers, catkins, and leaves can be eaten raw or cooked. They contain salicylic acid but in smaller quantities than the bark. The leaves can be harvested any time during spring and summer. The younger the leaf, the softer the texture. Their flavor is astringent and fresh. Young leaves can be eaten raw, but the older leaves are better dried, crumbled up, and added into baked goods like bread or muffins. The inner bark can also be eaten, but I have yet to try this. Some reports say it tastes sweet, others say bitter.

> **Caution**
>
> Do not use willow bark if you are allergic to aspirin. If you are pregnant or nursing, consult with your doctor before using. I have come across warnings against giving willow bark to children under sixteen who have the flu, chickenpox, or a viral infection. They can develop Reye's syndrome which is a very serious, albeit rare disease.

Other uses

Willow bark combats inflammation in the body and can be used as a painkiller. If I have a headache or an inflammation, I make a decoction or an infusion with the bark. If you are out in the forest, you can chew directly on a branch; it will have the same effect. It contains salicylic acid, which is what makes aspirin effective. It works less quickly than aspirin, but with none of the negative side effects on your gut.

I like to harvest my yearly supply of willow bark at this time of year, early spring, as the days are getting warmer. I try to use bark from the branches instead of the trunk; it does less damage to the tree. Willow bark can be harvested any time of year, but it is easier to do in early spring when the sap is running. I cut the harvested bark with secateurs into 1 to 3 cm chunks and dry them. I store them in glass jars in a dark and dry place. They keep their potency for about a year or two.

The branches can be used as a rooting hormone tea for when you take cuttings of berry bushes or trees. The salicylic acid and the indolebutyric acid in the willow branches make the other cuttings develop roots faster, and help them to fight off bacteria, fungi, and infection.

Figure 6.3. Top and bottom of leaf.

Figure 6.4. Bud.

Similar species / look-alikes

There are many different species of willow, and they are hard to tell apart from one another. There is also much hybridization within the willow family. A thorough description of them all is beyond the scope of this book.

- **Pacific willow** (*Salix lucida* ssp. *lasiandra*) Does not have drooping branches, grows in an upright fashion. Leaves are not as skinny nor as long as the leaves of the weeping willow. Pacific willow is a native tree that is an important winter food for moose. Can be used the same.

Willow basket weaving

Anneh Kessels operates **Willow Bender Baskets** and has been making baskets and teaching people how to work with willow for eighteen years. She is mostly self-taught and talks about how basket weaving has been part of every human culture for centuries. Our bodies know how to do this, she says, and her love and enthusiasm for weaving is infectious. She loves working with willow because it is forgiving; it lets you learn as you go. Willingness to learn from mistakes is important when weaving baskets. It might take some perseverance through the tricky parts, and it demands patience. The weaving can at times be hard on the hands, but Anneh laughs and says that she absorbs enough painkiller from the oils in the bark of the willow (which contains salicylic acid). Anneh travels throughout BC, teaching at various sites. At the end of each course, the students will have their very own basket to take home, and the know-how to make more baskets in the future.

Figure 6.5. Pacific willow.

Figure 6.6. Willow basket. Credit: Kristin Charleton.

Anneh uses more than twenty different types of willow, both wild and cultivated. With so many varieties, sometimes she cannot quite name which species she is using. For weaving purposes, you do not need to know the taxonomic name, but you do need to know if it is a type that is suitable for weaving. The common weeping willow is not ideal for this purpose. Anneh explains that it has a higher ratio of pith than wood in the branches and will not make a sturdy basket. It is better to choose a willow with more wood. To determine this, cut a branch and observe what it looks like in cross-section.

Anneh collects her branches throughout the winter. The sap should still be in the ground, and the shrub should be leafless. The branches range in thickness from the size of a pinkie to even smaller and are at least 4 feet (1.2 m) long, preferably even longer. Anneh sorts the willow branches to size, bundles them up, and dries them for several months. The fresh willow sticks shrink as they dry. Anneh explains that, if you weave with fresh sticks, your basket will become loose once the sticks have dried. There is a short period when you can use willow rods that are semi-dry. This is usually around five

Figure 6.7. Anneh harvesting willow in winter. Credit: Liliana Dragowska.

to six weeks, when they have dried enough to not have major shrinkage; not too dry that they break, not too fresh that they shrink.

She picks from the same trees from year to year. When a willow tree has been coppiced (had its trunk cut down close to the ground), it will sprout straight shoots each year. These are ideal for weaving. Willow can be harvested sustainably this way because trimming does not kill it, but rather makes it grow more vigorously.

Soak the dried branches before using. This is a fine art. If the time in the tub is too short, the willow will not be flexible enough; if it is in too long, the willow will become mushy. Anneh uses branches with the bark on. After soaking them for seven to ten days, depending on their size, you can check if they are ready by trying to bend the branch at a 90° angle. If it does not snap, it is done. The next step is called "mellowing the willow," which is letting the branches rest, wrapped in plastic or old sheets, for one to two days. After this, the weaving can begin. It is important to wrap the branches and the basket in something that will hold the moisture (e.g., plastic bag or tarp) when you take a break overnight. The usability of soaked willow branches is time-sensitive, so once you start weaving, you must complete your project within a few days. Under the right circumstances, the willow wrapped in plastic can be kept for weeks. It is not recommended to resoak the willow, as it then turns mushy.

When I asked Anneh what tools she uses, I was surprised to learn that the only essential tools are a knife, a pair of pruners, and a big rock/weight to hold the bottom of the basket she is weaving. Sometimes an awl can be useful to poke openings in the basket, but this tool can be exchanged for a larger willow stick. At times, a rapping iron is also used to gently rap the willow down for a tighter weave. No other materials are needed in a basket other than the willow branches. Anneh uses no twine, wire, or string.

Anneh also creates sculptures, gates, and fences from willow. She is hoping to create furniture from this incredible material soon too. Your imagination is the only limit when it comes to willow. To see more about Anneh's art and workshops, see willowbenderbaskets.weebly.com and her Facebook page WillowBenderBaskets.

Norway Maple

Acer platanoides | Sapindaceae

Synonyms

Harlequin maple

Description

Tree (30 m tall with a diameter of 1.5 m). Can reach an age of 250 years. Introduced from Europe as a landscaping tree. **Bark** is grey. Terminal **bud** is rounded and burgundy red. Petiole has a groove with three points to fasten it to the stem. When the leaf is snapped off, these three points **exude a milky fluid. Leaves** have five lobes with sharp points. Leaves are fluorescent green in spring, dark green in summer, and yellow in the fall. One of the last maples to lose its leaves in the fall. **Flowers** are yellow green with five rounded petals and five triangular sepals. They have eight stamens surrounding one yellow pistil. The flowers' corymbs smell sweet and attract hordes of pollinators.

Samaras (seedpods) are bright green in spring and brown in fall, often left hanging in the tree throughout the winter. They are shaped like wooden coat hangers. When ripe, the samara splits, and each winged **seed** flies its own separate way like a mini helicopter. The Norway maple is prolific at spreading its young. It can handle living in the shade.

Acer comes from Latin and means "sharp." The Romans used maple to make spears, which "sharp" refers to. The name *platanoides* refers to its likeness to trees in the genus *Platanus* (plane trees).

Figure 7.1. Leaf and samara.

Uses

The samaras are delicious when young and green. I eat the entire thing; it is a little bit sour and reminiscent of a pea pod. I prefer them fresh to cooked, since they lose their vibrant color when exposed to heat.

The young leaves are also edible but tend to be astringent. The mature leaves are not very tasty but can be dried, crumbled into powder, and added to soups or baking. The flowers have a unique flavor, both sweet and astringent. They are dainty and delicate but have enough crunch in them to feel substantial. I like to add them to salads, fruit salads, or sandwiches.

Figure 7.2. Milky fluid from petiole and stem.

Figure 7.3. Samaras.

The ripe seeds can be utilized if leached of tannins. Tannins occur naturally in red wine and tea and are not harmful to consume in small quantities. Ingesting too much tannin can give you a stomachache. I harvest my seeds around the same time as the tree loses its leaves. My leaching method is to boil them in a pot of water for ten minutes. The water turns dark purple. I strain them and repeat four or five times until the water is clearer and the seeds taste less bitter. I pan roast them immediately, adding salt. They taste great eaten just like that. They remind me of cooked lentils and can be added to meals or ground into flour and used in baking.

Figure 7.4. Bread made from dried, mature Norway maple leaves.

Figure 7.5. Flowers.

Tapping maple sap

Most people associate tapping maple sap with sugar maples (*Acer saccharum*), with good reason. You are more likely to get better-quality syrup and higher yield from those trees. However, if you do not have a sugar maple, any other maple will also give you tasty sap. The most

Figure 7.6. Seed.

Figure 7.7. Tapping Norway maple.

important aspect of tapping sap is timing. The weather should still be crisp, below freezing at night, and warm during the day. The sap starts to run as the ground thaws but before the leaves unfurl. You want your tree to be healthy and strong.

In order to tap sap, you need some equipment. We make most of our equipment from material we have on the farm, but there are sets to pick up at any outdoorsy/DIY store. They do make it easier, so if you are thinking of doing this every year, it might be worth the investment. When you drill into the tree, do it on a warmer day; this minimizes the chance of the wood splitting. I drill at a slight upward angle to the recommended depth of 6 to 6.5 cm. Remove any shavings, and the sap will flow immediately. Insert the spile (a spout that directs the sap into your collecting device) and tap it into the tree with a hammer. If it is too hot during the day, the sap can spoil if left in the bucket too long. As soon as night temperatures are continuously above freezing, the tapping season is over.

When you pull the spile from the tree, it will seep for a little while, but the wound heals quickly. You can plug it with clean beeswax or leave it to heal on its own. Our tree does not seem to have suffered any harm from having a tap. It has gone through the stages of flowering, setting seeds, and leafing out without any perceivable ill effects. Next year, we will tap a different one. This will give the tree a chance to regain any losses that were incurred by us.

The sap is sweet and thirst-quenching. We turned some of it into syrup. You need crazy amounts to get any syrup, so we only tried a little, just to see how it turned out. Some estimates say it takes sixty liters of Norway maple sap to get one liter of syrup. The sugar maple has a ratio of 40 to 1. It was tricky to cook it for the right amount of time to get the consistency we wanted. Just when we thought we almost had the right consistency, we looked away for a second...and it burnt to the bottom of the pan. After all that! We drank a lot of the sap fresh without any preparation at all, and it was delicious. It is jam-packed with nutrients and minerals such as potassium, calcium, and magnesium.

Caution

It is often barren under a Norway maple; the tree exudes chemicals to inhibit other growth. Some people react with red and itchy skin from the milky sap.

Similar species / look-alikes

Norway maple has a milky fluid in the petioles. All the species below lack this telltale sign.

- **Red maple** (*Acer rubrum*) More common in eastern North America. Leaf has three lobes. Samara is formed in a V rather than in a coat hanger shape. **Can be poisonous to horses.** See caution box on page 255 [re Rocky Mountain maple].
- **Sugar maple** (*Acer saccharum*) More common in eastern areas of Canada and the US. Leaf has five lobes. It is known for its fall color display in red, yellow, and orange. Samara has a square look. Edible.
- **Broadleaf maple** (*Acer macrophyllum*) Tall tree (up to 48 m). More common on the Pacific coast. Leaves have five lobes that are deeply incised and are much larger than Norway maple. Samara is V-shaped. Edible.
- **Rocky Mountain maple** (*Acer glabrum*) Native slender tree (up to 10 m tall). Leaves have 3 to 5 lobes with coarsely toothed edges. See Rocky Mountain maple chapter on page 253.

Kinnikinnick

Arctostaphylos uva-ursi | Ericaceae

Figure 8.1. Kinnikinnick shrub in July.

Figure 8.2. Flowers in mid-May.

Synonyms

Common bearberry, bearberry

Description

Hardy perennial shrub that grows close to the ground. **Leaves** are evergreen, leathery, and stiff, 1–3 cm long with smooth edges. They are teardrop shaped, wider at the tip than at the base. Leaves are bright green in the spring, dark green in the summer, and shift to reddish in the fall. **Stalk** is woody. **Bark** is greyish when older and reddish on new growth. **Flower** looks like an upside-down white urn with pink frills. It is small, 4 to 6 mm long, and can be seen between May and June. **Berry** is bright red when fresh and dark brown/red after a winter under the snow. The texture of the fruit is mealy. It contains five or six big brown **seeds** that are shaped like orange wedges.

Figure 8.3. Kinnikinnick berry and seeds.

Spring Urinary Tract Tonic

First make a cold infusion with the leaves. A hot infusion releases the tannins in the plant and makes a bitter beverage. Tannins are not necessarily bad (in small doses), but they make the beverage bitter. See caution box.

- ½ cup chopped fresh leaves
- Put in jar with cold water, cover. Let sit at room temperature for up to 12 hours. Strain out leaves.
- Add to liquid:
  - Pinch of baking soda (creates an alkaline environment)
  - 1 tsp. honey
  - A tiny splash of lemon juice (omit if you have a urinary tract infection as it might make the mixture slightly less effective at combatting UTIs).

Arctostaphylos means "bear grapes" in Greek, and *uva-ursi* means "grapes of the bear" in Latin. Bears eat the berries both for food and to tranquilize parasites within their body. Afterwards, they eat a particular sedge, and the fibers in that sedge bring out the tranquillized parasites. I find this fascinating! How did the scientists figure this out, and even more remarkable, how did the bears?

Uses

The flavor of the berry improves during winter. In March it is sweet and juicy. Not many berries remain on the shrubs at this time of year though, since they are popular not only with bears but also with rodents and birds. The frozen berries can be dug up from under the snow and eaten raw. The kids love this; it's like a treasure hunt in winter wonderland.

Other uses

The kinnikinnick leaves can be harvested throughout the year. They can also be dried and stored. The shrub appears to be thriving in BC, but in some other provinces and states, it is considered imperiled or vulnerable.

The leaves of kinnikinnick often have brown/black dots; this is a fungus that alternates between the kinnikinnick plants and spruce trees. It shows up as witch's broom clusters in the spruce trees. I try to avoid using the ones with the fungus.

Take great care while harvesting, and do not take too many leaves from any one bush. I like to use the leaves to make a tonic that tones the muscles and membranes in the urinary tract and helps to control the bladder. It can also prevent little kids from bed-wetting by helping them to completely empty the bladder before going to bed. I find that the leaves also help with constipation.

The tonic tastes delicious, and my kids really like it. We drink it occasionally with meals. For the kids, I limit their use to two days and three glasses in total. Then we might not drink the tonic for several months. The cold infusion can also be frozen or turned into delicious popsicles. To make them extra special, put an early blue violet (*Viola adunca*) in each. It is a treat both for the taste buds and the eyes!

Kinnikinnick leaves have been used in herbal medicine for a long time as a urinary tract disinfectant. It can help to treat uncomplicated, recurrent infections. It is antiseptic, anti-inflammatory, and antimicrobial. Herbalists sometimes combine it with urinary demulcents such as marshmallow root (*Althea officinalis*) and corn silk (*Zea mays*). Kinnikinnick works best in an alkaline environment. Therefore, sodium or potassium bicarbonate is often given together with the herb to create an alkaline environment conducive for the plant to work its magic.

Kinnikinnick leaves were and are an important component in smoking mixtures among the Secwépemc and Interior Plateau Peoples. They dried the leaves in the sun and then toasted them over a fire before mixing with tobacco. The name *kinnikinnick* comes from a Proto-Algonquian language and means "mixture."

Caution

Kinnikinnick leaves contain arbutin and tannins and can cause stomach and liver problems if used too much. Some reports say to limit intake to two weeks, and others say three days. Always err on the side of caution, especially with children or frail individuals. It should not be used by pregnant women because it can stimulate uterine contractions and make the placenta uterine membrane more permeable. The use of the leaves might turn the urine slightly green, which is no cause for concern, but if you are giving a urine sample, let your doctor know ahead of time so as not to cause alarm.

One way to encourage further growth of the plant is to lay a stem down and cover it with soil (layering propagation). In this way, the plant forms an extended ground cover.

Similar species / look-alikes

- **Alpine bearberry** (*Arctostaphylos alpina* var. *alpina*) Leaves are more veined, and berries are black. Grows at higher altitude. Edible.
- **Red bearberry** (*Arctostaphylos alpina* var. *rubra*) Leaves are more veined. Berry is red and juicy rather than mealy. Grows at higher altitude. Edible.

Figure 8.4. Fall colors.

Early Blue Violet

Viola adunca | Violaceae

Synonyms

Western dog violet, hooked spur violet, Kirk's violet, western blue violet, western long-spurred violet, hooked violet, sand violet

Description

Perennial. Early blooming plant (2 to 8 cm) in shady spots with dappled sunlight. Grows from a **rhizome**. **Flower** has five petals: two at the top and three at the bottom. Toward the center, the petals are white with black streaks. The two bottom petals on either side have white beards. Pistil is yellow/orange. Lowest petal has a **spur** that is about half the length of the bottom petal. Purple/green **stalk** has fine hairs. Flower and leaves grow on separate stalks. The flower stalks stick up higher than the leaf stalks. Flower gives off a very strong violet scent. **Leaves** are heart shaped with jagged edges (1 to 2.5 cm wide). They are covered in very fine hair. In midsummer, a mottled white and purple **seedpod** forms close to the ground; when ripe, it catapults its seeds. Violet **seeds** are coated in a special oil that ants find irresistible, and they carry the seeds with them underground. They eat the fat surrounding the seed and then leave the seed to grow. Perfect for the violet, which is now planted in a protected place.

Viola refers to the violet genus. *Adunca* means "bent toward" or "hooked."

Important fact!

Viola adunca is hugely important for some Zerene fritillaries butterflies (*Speyeria* spp.). Their larvae feed solely on early blue violet. They also pupate in the violet debris. Without this plant, they will no longer exist, and many of the species are on the endangered list. These violets should be encouraged to take over our

Figure 9.1. Early blue violet.

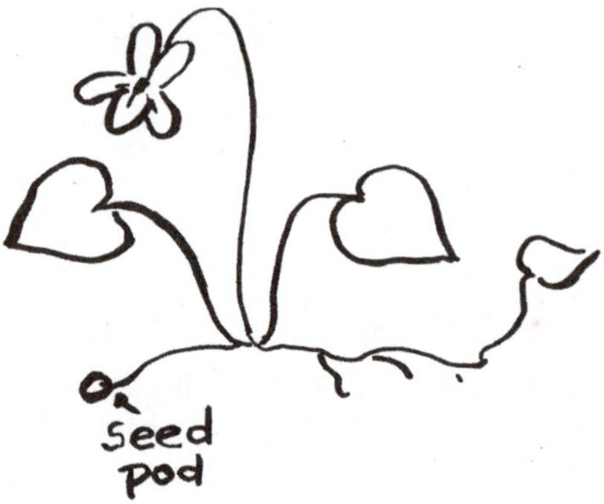

Figure 9.2. Note location of seedpod.

lawns to give these beautiful butterflies a chance of survival. If you wildcraft this species, do it gingerly and make sure that any critters in and amongst the violets do not get injured. Also, check the status of the plant where you are and do not pick if it is threatened.

Uses

There is no other smell that can compare to a patch of blue violets in early spring. It makes my mouth water! It reminds me of violet-flavored candy I ate as a child. We use the flowers and the leaves in a variety of ways. Flowers placed in the sugar bowl will impart that delicious flavor of violet to the sugar and can then be used in baking. Make sure you use no other overpowering flavor or scents as the violet flavor and scent is quite delicate and can easily be lost among other stronger flavors. We eat the flowers raw or put them into popsicles or ice cubes as well.

I also solar-infuse water with the flowers and use this liquid as a drink. I place my solar infusion on the kitchen window ledge for a day max, and then I strain the flowers and add them to baking. The remaining liquid will have captured the ephemeral scent of the violet. It can be frozen for use later in the year. The smell is exactly as the day the violet bloomed when you open that jar. I sometimes keep a small jar in the fridge. I open it and take one big gulp, and I can both taste and smell early spring. It makes me happy!

As beautiful as the flowers are, I appreciate the leaves even more. Not so much in their raw form, even though they can be eaten like that,

Figure 9.3. Intact early blue violet seedpod and seedpod after "catapulting."

Caution

Violet is an important habitat for Zerene butterfly larvae. See important fact on page 17. Use this violet in moderation due to saponins in plant. If eaten in too great a quantity, violet can cause digestive upset. Eat only the leaves, stems, and flowers. The roots and seeds are toxic and can cause intestinal upset and possible respiratory and circulatory problems.

Figure 9.4. Side view of flower showing spur.

but as a cooked green. They start to grow when there are few fresh greens to be had, and they work well to thicken soups and stews. As part of cooked greens in a lasagna, they shine. I also like them in smoothies. In March, we pick the leaves, but we need to wait until April before the flowers appear. Early in the spring, before flowering, they can be hard to tell apart from other violets. Luckily, the others are also edible. Sometimes the leaves of wild ginger (*Asarum* spp.) are confused with early blue violet. See page 20.

Other uses

Violet is loaded with vitamins A and C. The entire plant is an emollient, meaning it has a softening effect on the skin and can be used as a poultice or in skin creams. A tea made from the leaves is said to be slightly laxative and to aid in sleep. We have not used this plant specifically for medicine or for external use, so I

Figure 9.5. Early blue violet leaf top and bottom.

Figure 9.6. Wild ginger left, violet right.

cannot attest to how well it works; considering its mucilaginous quality, I imagine it would work well.

Similar species / look-alikes

- **Marsh violet** (*Viola palustris*) Flower is pale purple or white. Spur on flower is much smaller. Leaves are wider. Edible.
- **Alaska violet** (*Viola langsdorfii*) Does not have bearded styles. Spur on the flower is shorter and thicker. Edible.
- **Wild ginger** (*Asarum caudatum*) (*Asarum canadense*) Flower has three maroon sepals. Leaf lacks serrated edges. Edible. See wild ginger chapter on page 2.

Cleavers

Galium aparine | Rubiaceae

Synonyms

Catchweed, sticky willy, robin-run-the-hedge, clivers, goosegrass, couchweed, Velcro plant, catchweed bedstraw, stickyweed

Description

Perennial. Grows like a vine. **Stalk** is square with four ridges and sticky hairs. **Leaves** are six to eight in a whorl. They are narrow and have a sharp point at the end with a single distinct vein. Underside of leaf has hooked hairs. Leaf edges have sharp bristles that point backwards. Sweet-scented bedstraw (*Galium trifolium*), a close look-alike, has bristles that point forward. Small **flowers** have four white/yellow petals and form on stalks (peduncles) from the leaf axils. **Seeds** are paired with hooked hairs. The entire plant sticks to fur and clothing. When drying, the plant emits a sweet smell due to a chemical called coumarin.

The name *Galium* comes from *gala*, a Greek word for "milk." It was given this name because of the plant's ability to curdle milk. *Aparine* comes from the Greek *apairo* which means "to seize or grab".

Uses

In April, young greens are just starting to poke out, and it is such a joy to be able to harvest fresh greens again. The young cleavers are soft and delicious at this time of year. So far, they have only grown 5 cm tall, and the entire plant

Figure 10.1. Note cleavers square stem and leaves in a whorl.

Figure 10.2. Young plant.

Figure 10.3. Cleavers chaff and seeds to be ground.

Figure 10.4. Flower stalk from leaf axil.

can be eaten raw. As soon as they develop too many prickly hairs, I cook them instead. They taste great in a stir-fry or eaten as cooked greens. When it gets too old and stringy, the entire plant can be made into a delicious sweet tea with undertones of vanilla. I strain it through a fine mesh to avoid the little hairs. To store it, I dry the entire plant thoroughly and keep it in a dry and dark place. If the plant accidently molds, the coumarin in it converts into dicoumarol, which is a powerful anticoagulant compound. If the plant has been dried and stored properly, it poses no such risk. See caution box.

Figure 10.5. Cleavers flower in July.

Figure 10.6. Seeds/nutlets in August.

Later in the year, the small nutlets/seeds can be harvested, roasted, and ground to make a coffee substitute that contains no caffeine. Cleavers is in the same family (Rubiaceae) as the coffee that we regularly consume. It is very similar to coffee but tastes more earthy. It tastes

Figure 10.7. Cleavers coffee.

Caution

Contains coumarin. Be careful to store this plant properly and do not use it if it is moldy. The coumarin in the plant changes into dicoumarol during the molding process and if ingested acts like an anticoagulant. There is a heightened risk of excessive bleeding from any bruise or cut if dicoumarol is ingested. People on prescribed blood thinners should consult their doctor before using. Contains asperuloside, which stimulates the uterus. Do not use if pregnant.

Cleavers can easily be mistaken for sweet-scented bedstraw (*Galium triflorum*) which is considered at risk or high risk of extirpation in some areas. See look-alike section to tell them apart.

Figure 10.8. Northern bedstraw.

amazing! Roast the seeds until they are brown and crispy, about 20 minutes at 350°F (175°C). The flavor changes dramatically during roasting, and the aroma is delectable. The seeds are fiddly to harvest. To make it more worthwhile, the entire plant, including the stems and leaves, can be dried and roasted, and then ground and brewed.

Other uses

Cleavers have been used by herbalists as a diuretic and to help with skin conditions such as psoriasis and eczema. I have not used it for medicine.

A close relative, sweet-scented bedstraw (*Galium triflorum*) gives away its former use as a stuffing of matresses. It has a springiness to it, and it smells sweet when it dries. In addition, the plant can be useful as a makeshift sieve if you layer it crosswise.

In our family, we play Galium tag. Here is how you play it: You try to attach a strand of sticky Velcro-like Galium to the other players' clothes. Once the Galium sticks to their clothing, you run off so that they cannot stick it back on you. It is great fun!

Similar species / look-alikes

- **Sweet-scented bedstraw** (*Galium triflorum*) Looks almost identical. To tell them apart look at the bristles on the leaf edges. In sweet-scented bedstraw the bristles point forwards and in cleavers the bristles point backwards. The leaves are more elliptical and are in whorls of 4–6. Smells stronger of vanilla. Can be used in the same way.
- **Northern bedstraw** (*Galium boreale*) White flower clusters on top give it a different profile. Grows erect rather than as a vine. Leaflets are arranged 4 in a whorl. Edible.
- **Bedstraw** (*Galium verum*) Flower is yellow and grows in clusters at top. Grows in an erect manner. Leaves are skinny, 6 to 8 in a whorl. Edible.

Purple Dead Nettle

Lamium purpureum / Lamiaceae

Synonyms

Red dead nettle, bumblebee flower, purple archangel

Description

Annual. Early spring plant (up to 30 cm tall). Grows in disturbed areas with lots of sun exposure where it often forms great stands. Creates adventitious **roots** from each node that touches the ground. The soft and hairy **leaves** are triangular with jagged edges and net-like veins. Lower leaves are green, top leaves have a purple hue. **Stalk** is square. Lacks the minty scent of its relatives. **Flowers** are pink/purple and white with darker pink patterns. They are tubular and covered with fine hair. **Seeds** are grey with a meaty white tip and are held in jagged cups; each cup contains four seeds. Purple dead nettle and ants have a mutualistic relationship. The seeds have a coating of elaiosome that is rich in protein and fats. Ants carry the seeds underground and feed the lipid-rich elaiosome to their larvae. By taking the seed underground they help the plant to propagate.

The name, purple dead nettle, gives it an undeserved sinister air. The name comes from its resemblance to stinging nettle (*Urtica dioica*) but tells us that it does not sting. *Lamium* comes from the Greek *lamia* meaning "gaping mouth" and refers to the flowers. *Purpureum* means "purple."

Uses

I like harvesting purple dead nettle at this time of year because it grows so abundantly. The plant's leaves appear in large numbers in early spring, but it will not flower until warmer weather. They germinate very early and are one of the first weeds in our spring garden. The entire plant can be eaten raw or cooked. I find it a bit too fuzzy to eat great quantities raw. As a cooked green, it tastes lovely. I like them sauteed in oil with a bit of lemon juice and parmesan sprinkled on top. They do well as fillings in quiche, filo pastry, and lasagna. Their flavor is like spinach but sweeter. The flowers can be

Fig 11.1. Purple dead nettle flower.

Figure 11.2. Seeds.

Figure 11.3. Young purple dead nettle.

Figure 11.4. Stand of purple dead nettle.

eaten raw or added to salads. They are loaded with sweet nectar. The entire plant can be used as a tea, both fresh and dried. I prefer to make tea from the plants when the flowers emerge, which add a nice sweetness. Purple dead nettle is high in iron, vitamins, and fiber.

Other uses

The plant is rich in tannins which can help heal wounds by tightening up the skin. It has also been prescribed by herbalists as a tea to stop internal bleeding and to alleviate diarrhea.

Figure 11.5. Purple dead nettle left and henbit right.

Similar species / look-alikes

- **White dead nettle** (*Lamium album*) Flowers grow in distinct rings around the stalk. Has triangular leaves that are pointier than purple dead nettle. It looks more like stinging nettle. Edible.
- **Henbit/henbit dead nettle** (*Lamium amplexicaule*) Leaves have no stalks. Leaves clasp the stem. Leaves are fan shaped rather than triangular. Edible.
- **Stinging nettle** (*Urtica dioica*) Has stinging leaves. Flowers are greenish yellow and much smaller. Edible. See stinging nettle chapter on page 36.
- **Catnip/catmint** (*Nepeta cataria*) Soft blue-green leaves. Emits strong musky/minty scent. Flowers are white. Edible. See catnip chapter on page 58.
- **Ground ivy** (*Glechoma hederacea*) Grows prostrate and sprawling. Leaf is round and kidney shaped with scalloped edges. Purple flowers grow in whorls from square stem. Edible.
- **Hedge nettle** (*Stachys palustris*) Taller (up to 80 cm). Opposite pairs of hairy leaves. Leaves are longer and more narrow than dead nettle. Edges are more finely serrated. Purple flowers form on spike at top. Edible.

Oxeye Daisy

Leucanthemum vulgare | Asteraceae

Synonyms

Marguerite, moon daisy, common daisy, field daisy, poor land daisy, moon penny, poverty weed, dog daisy, horse daisy, maudlin wort

Description

Perennial herb (up to 80 cm tall) that likes full sun. **Flower** has white petals (ray florets) that surround the yellow tubular disk florets clustered in the middle. It looks like a small white sunflower. **Leaves** are paddle shaped with rounded lobes (spatulate). They grow both as a basal rosette and in an alternate pattern up the stalk. Further up the stalk, the leaves clasp the stem and are smaller and linear-oblong. The petioles taper off to a skinny point. Leaves are dark green and shiny. The green/yellow **stalk** is five-sided and slightly woody. **Root** is a strong **rhizome. Seed** is a round black achene without pappus, up to 1 mm in length.

The white ray florets function as a billboard to attract pollinating insects such as bees, flies, butterflies, moths, and beetles. If it is not pollinated by external sources, it can self-pollinate by bending the stigma to meet the pollen-loaded anthers.

Figure 12.1. Note the green bracts below the oxeye daisy petals.

Figure 12.2. White ray florets surrounding yellow tubular disk florets.

Figure 12.3. Basal leaf is spatulate; note the rounded lobes and long petiole.

Leucanthemum means "white flower" from the Greek words *leukos* ("white") and *anthemon* ("flower"). *Vulgare* is Latin for "common."

Uses

The entire plant is edible and tasty. The basal leaves form a rosette and are delicious. They have a nice crunch to them; the flavor is sweet and slightly perfumed. Their appearance is beautiful in a salad, and they come in a perfect bite size. I can eat a salad composed of only these petals and feel satisfied. Raw, they are lovely on sandwiches, and in smoothies. I find them refreshing when combined with sheep sorrel (*Rumex acetosella*). They can also be cooked and added to omelets and stir-fries. We harvest them throughout the season, from April to September. The leaves remain juicy and soft even when the plant is flowering.

The flower can also be eaten. It tastes great fried in butter or made into a tea. Some sources mention wine made from the flower heads, but I have yet to try this. The unopened flower buds can be eaten raw or be pickled to make a caper-like condiment. The seeds can be eaten. If you find a large stand, they can be worthwhile to gather. I like to sprinkle them on yogurt.

Figure 12.4. Oxeye daisy flower bud.

Figure 12.5. Stalk and cauline leaves (cauline—attached to stem).

Other uses

Tea made from the flowers has been used to relieve chronic coughs and bronchial catarrhs and to clean up sinus infections. The plant is picked at the height of summer and then dried and stored. Some people use oxeye daisy in similar ways to chamomile to calm down an overexcited mind. It can also be applied externally to heal wounds.

> **Caution**
>
> Can cause contact dermatitis in some people. Some reports say to not use if pregnant or breastfeeding, and to avoid if you have kidney problems.

The plant contains pyrethrum, which is a natural pesticide. The dried flowers have been used to deter fleas. When cows eat the plant, their milk comes out tasting unpleasant to humans.

Similar species / look-alikes

- **Shasta daisy** (*Chrysanthemum* x *superbum*) Cultivated and larger. Leaves are linear with jagged edges. Edible.
- **German chamomile** (*Matricaria chamomilla*) Leaves are fine and feathery rather than paddle shaped. Edible.
- **Roman chamomile** (*Chamaemelum nobile*) Leaves are fern-like. Edible.
- **Mayweed** (*Anthemis arvensis*) Leaves are fern-like. Used medicinally.

Hairy Bittercress

Cardamine hirsuta | Brassicaceae

Synonyms

Spring cress, land cress, lamb's cress, hoary bitter cress, shot weed, flick weed

Description

Winter annual or biennial (up to 30 cm tall). Native to Eurasia. Likes full or partial sun and soil that has been disturbed. Can grow in most soils, even gravel. Starts as a basal rosette. **Leaves** are simple-pinnate with 5 to 9 leaflets. Petiole is long, green to purple with hairs along edge. Each leaflet is round to oval and wavy along the edge. It is often hairy. Basal leaves are as long as they are wide. In the middle of the basal rosette, a flower stalk will shoot up. Cauline leaves alternate up the stalk, and the leaflets are

Figure 13.1. Entire hairy bittercress plant.

narrower. Stalk terminates in a raceme of flowers. Tiny white **flowers** (2 mm) bloom at the top. The flower petals form a cross; this is how plants of this family became known as Cruciferae. Hairy bittercress has four petals, four sepals, four stamens, and a single styled pistil. The green-purple sepals are shorter than the petals.

Once it has been fertilized, it forms narrow cylindrical **seedpods** (siliques). They are brown/purple at maturity and lack winged margins. Inside are several small ovoid **seeds** that are slightly flattened. At the slightest touch, the ripe seedpod explosively splits in two and scatters its seeds a far distance. This is called ballistic seed dispersal or ballochory. It is believed that small bees and flies pollinate hairy bittercress. **Root** is shallow and fibrous.

The falcate orangetip butterfly (*Anthocharis midea*), the spring azure butterfly (*Celastrina ladon*), and the purple-backed cabbageworm moth (*Evergestis pallidata*) occasionally feed on bittercress. The name *Cardamine* comes from the Greek *kardamon* and means "cress," *hirsuta* means "somewhat hairy."

To be perfectly honest, I have a really hard time telling this bittercress apart from some of the others in the area. Small-flowered bittercress (*Cardamine parviflora arenicola*) and little

Figure 13.2. Flower.

Figure 13.3. Leaf.

Figure 13.4. Hairy bittercress seedpods (siliques).

western bittercress (*Cardamine oligosperma*), in particular, are alike and are described similarly in the literature. Despite this confusion, I feel comfortable identifying the plant as a bittercress, and I have liked all the bittercress I have tried. To be sure that I do not accidentally harm one of the threatened species, I make sure to spread the seeds of any plant I harvest and to keep the habitat intact.

Uses

These are welcome plants in my garden and put a smile on my face when they pop up in late winter/early spring. They will seed in late summer and start to grow in late fall. The snow and cold usually puts a stop to them before they get very far, but as soon as the snow melts and spring is back, they continue growing with vigor and pop up everywhere in my dug garden beds. I help them along by scattering their seeds where I want them to grow. I love the flavor of this plant. It is spicy, but not too spicy. It tastes a bit like mustard greens but is less pungent. It is reminiscent of cilantro mixed with arugula. I love it on a sandwich or with egg salad on fresh bread. It can also be cooked. I eat the entire plant. Since it is so small, it is hard to get great quantities. Make sure to wash the plant well, as it tends to be quite sandy. If it is very mature, the stems and flowers can be a bit fibrous, in which case, use the leaves only. The root is woody, but it can be grated onto food.

The small seeds and seedpods are edible. The seeds can be ground up with other wild mustard seeds to make prepared table mustard. The ground seeds are more pungent and spicier. This plant can be grown for microgreens in winter.

Other uses

This plant contains antioxidants, vitamins A and C, calcium, magnesium, phosphorus, and beta-carotene. It is a stimulating herb that can be used as an expectorant. It also has some diuretic properties and is claimed to remove carcinogens from the body. I have not used it as medicine myself.

Mustard seeds are helpful in poultices/plasters. The seeds of bittercress are too small in quantity to make enough on their own, but they can be added to other wild and domestic mustards. A mustard poultice is made from mustard seeds ground into a fine powder. To that you add a little bit of flour and water to make a paste. I spread this on a cotton cloth and then apply the cloth to my chest when I have a cold or cough. The mustard poultice should not be placed right against the skin as it can cause blisters. After a few minutes on your chest, it warms up and feels wonderful. It can also be used on a sore back or sore muscles. Make sure you keep an eye on it and remove if skin starts to blister. A little reddening is normal. Be extra vigilant if you are using the mustard poultice on a child, as their skin is more sensitive.

Caution

There are many different species of bittercress, some of them are critically endangered, threatened, or vulnerable due to habitat loss. Make sure that you check the status of the bittercress in your area before harvesting. Consuming too much of this plant can give you heartburn.

Figure 13.5. Shepherd's purse basal rosette.

Similar species / look-alikes

- **Butterweed** (*Packera glabella*) Leaves are similar at first glance. Butterweed has leaves that have jagged edges, and the end leaf is larger than the rest. Easiest way to tell them apart is by the flowers. Butterweed has yellow daisy-like flowers with 5 to 15 ray florets. Do not ingest, as it is **toxic.**
- **Little western bittercress** (*Cardamine oligosperma*) Native to North America. Looks very similar. Leaves are less rounded and have 3–5 lobes. Flowers have six stamens, instead of four stamens like hairy bittercress. Can be used in exactly the same way.
- **Pennsylvania bittercress** (*Cardamine pensylvanica*) These two plants are hard to tell apart. This species lacks hairy petioles and has less hair on base of stalk. The end leaflet sometimes has three lobes. Other than that, they are virtually indistinguishable. Pennsyl-

Figure 13.6. Shepherd's purse seedpods.

Figure 13.7. Watercress.

vania bittercress tends to be bigger. Can be used the same way.

- **Shepherd's purse** (*Capsella bursa-pastoris*) Often grows in similar habitats. It has seedpods shaped like small hearts or like purses that shepherds used to carry, hence the name. It forms a basal rosette with lanceolate, pinnately lobed leaves. Flowers are white and like bittercress. Edible.
- **Watercress** (*Nasturtium officinale*) Rapidly growing aquatic plant closely related to the *Cardamine* genus. Looks very similar to bittercress, but it needs an aquatic environment. Flowers are small and white. Seedpods (siliques) are long, narrow, and erect. Leaves are alternate and pinnately compound. Leaflets are arranged in opposite pairs with a larger leaflet at the tip. Leaflet edges are wavy. Edible. Can host harmful parasites such as giardia and liver flukes when growing in contaminated water. **Pick with caution.**

Stinging Nettle

Urtica dioica | Urticaceae

Synonyms

Nettle, common nettle, tall nettle, American stinging nettle, California nettle

Description

Perennial herb that prefers moist and shaded areas. **Cord-like stem** with yellow **stinging hairs. Leaves** are attached in a pattern called opposite decussate, meaning that the leaves are in pairs at right angles to the pair above and below them. When the leaves are young, they have a purple tint. Later in the year, they are a rich dark green. Leaves have stinging hairs. I find that the hairs on the bottom of the leaves have more sting than on the top. **Flowers** are small, greenish white, and easy to overlook. Plant is pollinated by wind; needs both male and female plants to produce seeds. The small **seeds** are oval and green, turning brown later in the fall. **Root** is fibrous, and the stinging nettle spreads by **rhizomes.**

Urtica means "stinging," and *dioica* refers to the fact that male and female flowers are on separate plants.

Figure 14.1. Young stinging nettle plant.

Uses

We have been eating nettle for a while already. You can find them as soon as the snow disappears. Today we harvested stinging nettle for soup. I usually only pick the top two nodes to get the best quality and the most tender part of the plant. Nettle soup is one of my family's favorites. It is sooo easy to make.

I try to pick as many nettle leaves as I can squirrel away for the winter months. If you only pick the top of the plants, they bounce back and produce even more growth. Many jars of dried nettle are needed to serve our family throughout the winter. They will be used in soups, casseroles, bread, and waffles; added to omelets and smoothies; and used for tea. I feel that one can never have too much nettle. When harvesting the leaves, I pick them before the plant flowers. Some reports say that older leaves can be upsetting for the kidneys. The plant contains iron, vitamins A and C, and is easily digested by the human body.

The kids and I snack on raw nettle leaves while we are out in the woods. To avoid any stinging in our mouths, we fold them over with the underside of the leaf inside. We gently rub the leaves to dislodge the hairs and then pop in our mouths. My kids call them nettle pillows. They taste great. If you want to add nettle leaves raw to your diet, any friction will remove the stinging hairs. I find that ground-up raw leaves with goat cheese makes a fabulous cracker spread.

Later in the season, the seeds can be collected and eaten. They are used by herbalists to increase energy. Nettle seeds raise dopamine levels. See caution box. They give me a rush, like too much caffeine, so I consume them in small doses. A light sprinkling on my morning yogurt or a few thrown into bread is the perfect amount for me. They can be mixed with other seeds and nuts for a high-powered energy snack.

Figure 14.2. Seeds in July.

Figure 14.3. Close-up of stem with stinging hairs.

Other uses

Nettle is very helpful for people with anemia. It is high in iron and functions as a blood builder. Tea made from the leaves has been used as a diuretic and as a cleansing tonic. If you start using nettle early in the season, it can alleviate seasonal allergies to pollen. It gives relief without the side effects, like drowsiness, that are common with over-the-counter antihistamines. Even the stinging part of the nettle has been used. When the small needles/hairs touch your skin, they inject your body with acetylcholine and histamine.

Nettle Soup

- Chop nettles. Throw in pot with stock. Add some salt, pepper, and lemon juice.
- Cook for approximately 10 minutes.
- I use a staff mixer to puree it. It tastes delicious just like that.
- If you add fried onion and bits of bacon at the end, it turns into a decadent treat.
- I usually serve mine with half of a hard-boiled egg floating in the middle. Bread on the side is a bonus.

Tonight's loaf was a dandelion petal and lentil bread. Together with the soup, it left us satisfied. The soup is invigoratingly green! Full of chlorophyll.

Figure 14.4. Nettle soup.

This creates a small inflammation on your skin. The body then sends more blood to the area that has been touched, and blood circulation is thus improved. I love touching nettle with my hands. I have poor blood circulation in my hands, and they are often cold. After touching nettle, I feel a warm tingly glow in my hands that can last for hours. Nettle has been used to whip body parts that are in pain or suffering from arthritis and so increase blood flow and lower pain. This practice is called urtication. If you have sensitive skin or dislike the tingling sensation, I advise you to wear gloves when you harvest nettles.

The nettle also has an important role as a plant food for my garden. I place bundles in a barrel with water and stir them occasionally. This brown sludgy liquid becomes plant fertilizer for my garden beds. You can use the leaves for fertilizer any time throughout the year.

Stinging nettle is an important plant for many insects, and several lepidoptera caterpillars rely on nettle as their only food source.

The fibrous stalks of the nettle in late summer/autumn can be turned into a string/cord. The fibers are long so this can be done satisfactorily by a novice like me. It is remarkably strong, and it is handy to know how to produce your own rope should you need it. I dry the stalks until I need them. Before making rope, I soak the stalks in water for a day to make them less brittle. Then I pound them gently with a wooden mallet. This breaks up the fibers and allows me to access the soft, stringy part of the stalks. It is important to remove any hard parts. The stringy fiber that is left, I roll on my leg. I like wearing jeans for this activity. The fibers will start to form into a cord. By adding several strands together, the rope's strength will improve. The longer the fibers, the stronger the rope will be, but you can splice in smaller sections too. I have also made rope from fresh nettle stalks; this is serviceable but does not last as long as the dried stalks.

The late autumn stems can also be turned into paper. This requires the use of lye to soften up the fibers. Lye can be quite caustic, so I experimented making paper without it. I had to cut up the boiled stems very finely, and still it was quite stringy. I had better luck making paper with the dried nettle leaves. First, I mixed old paper scraps and dried leaves with hot water until it was a pulp. I strained the pulp through a screen, then placed it on a drying board to be sponged dry. After it had dried flat for a day or two, we had our own handmade paper. The more texture it had, the more I liked it. If you prefer a smoother finish, run your pulp through a blender before straining. The paper we made will be turned into greeting cards. If you want to be creative, add seeds. When you tire of the card, it can be planted in the garden.

Caution

Can interfere with allopathic drugs for hypertension and diabetes. Can also give adverse effect if mixed with morphine and/or alcohol. Nettle seeds can be over-stimulating in doses over 30 grams per day. They can prevent you from sleeping. Nettle plants can irritate the skin when touched, and some people have severe reactions to the stinging hairs. Interestingly, the juice from the nettle has been found to help against the sting. I used nettle tea while I was pregnant, and I have seen many reports of its excellent benefits for pregnant women due to its high iron content. However, I have also seen cautions against using it while pregnant, so I advise you to do your own research and consult your doctor to find out what works for you.

Figure 14.5. Stinging nettle rope.

Figure 14.6. Stripping away the outer fibers, the part that is used to make twine/rope. Credit: Joanne Cannon.

Figure 14.7. Dried nettle paper with varied textures.

Similar species / look-alikes

- **Catnip** (*Nepeta cataria*) Leaves are hairy and soft to the touch. Square stem. Smells minty. Edible. See catnip chapter on page 58.
- **Purple dead nettle** (*Lamium purpureum*) Square stem. Soft leaves, no stinging sensation. Purple/pink flowers. Edible. See purple dead nettle chapter on page 25.
- **White dead nettle** (*Lamium album*) Square stem. Soft leaves, no stinging sensation. Edible. See page 27.

Dandelion

Taraxacum officinale | Asteraceae

Synonyms

Lion's teeth, pissabed

Description

Perennial (up to 70 cm). Deep **taproot** that smells bitter. **Leaves** form a basal rosette. They are jagged and hairless with a fleshy midrib and a narrow base. Petiole of the leaf when cross-cut forms a hollow half circle. **Stalk** is hollow, smooth, and ranges in color from yellow to green to purple. Exudes a milky sap when cut. On top of each stalk sits one composite **flower**. It has bright yellow ray florets and no disk florets. Involucre consists of two rows of bracts; the outer row is bent backwards. Flower head turns into the characteristic **seed head**. Each seed is attached to a pappus with hairy white bristles that enable the seed to travel far and wide, which delights me and children of all ages. Any chance I get, I spread those seeds.

The dandelion can reproduce by parthenogenesis (without the aid of fertilization and pollination) if it needs to. It then becomes a clone of itself. Otherwise, it is pollinated by various insects. *Taraxacum* is a medieval name for "bitter herb." *Officinale* means that the plant has been used as medicine throughout history.

Uses

This is one of the best plants I know! This time of year, we pick the new fresh greens. We add them raw to our salads or our sandwiches. We also cook them in omelets, stir-fries, and soups. I dry as many as I can because they do not get any better than this. The leaves can be dried and used throughout the spring and summer, but they are particularly tasty early in the year. I use the dried leaves in baking, tea, and casseroles throughout the year.

When the flowers appear en masse, both our family and the bees rejoice. Usually, they are in full force come May. One of our favorite family meals is pan fried dandelion flower heads, and

Fig 15.1. Cheerful dandelion.

Figure 15.2. Close-up flower.

Figure 15.3. Dandelion (*left*) is often confused with chicory (*right*). See look-alike section.

Figure 15.4. Dandelion seeds.

they are so fun to pick and make with the kids. Easy and delicious! They can also be used to make dandelion quiche.

Dandelion jelly/syrup can be made from the petals and sugar. It has a delicate yet specific flavor. If you have ever had honey made from dandelion, you know what I mean. The bees rely on dandelion in a big way. This is one of the first flowers to come in great quantities when they are in desperate need of a good meal after winter to restock their hives.

I sometimes collect the dandelion seeds to sprout during winter; they are a welcome addition to our diet. In order to avoid any mold, the pappus need to be removed prior to sprouting them. They can also be grown as microgreens. In the past, I have even dug up the roots in the fall and had them growing in a pot on the table for easy-access nibble greens. If the leaves are shielded from the light, they become paler and less bitter.

Even the stalk is edible when harvested at the right time. I pick them from shaded areas where the plant has had plenty of moisture and tends to grow taller. I boil them in salted water for a minute or so and call them dandelion spaghetti. They are nice mixed in with egg noodles. Fresh, they make fun drinking straws too.

The root can be harvested at any time during the year. I eat some of it raw early in the spring, when it is less bitter than in summer or

Pan-fried Dandelion Flowers

- Pick flower heads (petals and involucre).
- Rinse with water.
- Coat with flour.
- Fry in butter or coconut oil.
- Salt lightly.
- Eat warm and enjoy! Even the most intrepid dandelion-hater cannot dislike this treat.

fall. I enjoy adding the root to stews, casseroles, and soups. Later in the year, I dry the root and use it for tea or as a part in "herbal coffee."

Peoples' strange dislike for dandelions in their lawns has always puzzled me. Here is tasty food with medicinal qualities that looks pretty and cheerful. It also helps the bees that we rely on for almost all our grown food crops. Why do some people then insist on spraying these lovely flowers with poison or dig them out at first glance to replace them with inedible lawn, usually with a non-native grass species, that they then have to mow, water, fertilize, and mow and mow and mow? What for? I'd rather keep the dandelions that need no maintenance, and which give me plenty of food throughout the year. According to the USDA National Nutrient Database, dandelion leaves have more vitamins A, K, and iron than regular spinach.

Other uses

The bitterness of the plant has its medicinal uses. It is helpful to eat before a meal since it makes more saliva and gets the bile juices flowing, which helps to break down food more

Figure 15.5. Root.

Figure 15.6. Dandelion flower pie.

Caution

Avoid use of this plant if stomach and bowels are irritable since it can aggravate the condition.

Figure 15.7. Leaf comparison (from *left* to *right*) prickly sow thistle, prickly lettuce, dandelion, and chicory. For flower comparison see prickly lettuce chapter on page 49.

effectively and so aids in digestion. The plant is helpful as a detoxifier of the body, and that is why I like to eat some of the raw root in the spring. This cleanses your liver and kidneys and kickstarts a sluggish digestion. Just be prepared to run to the bathroom a bit more. The leaves are a diuretic, meaning they draw out liquid from the body, and this means more bathroom breaks too. Dandelion is rich in potassium and magnesium and works brilliantly as a diuretic since it adds important minerals that are often lost with conventional diuretics. Dandelion has been employed in herbal medicine for centuries.

Similar species / look-alikes

- **Chicory** (*Cichorium intybus*) Leaves look very similar but are slightly less jagged than dandelion and have fine hairs on underside of leaf. Chicory has blue flowers. Edible. See chicory chapter on page 163.
- **Sow thistle, perennial** (*Sonchus arvensis*) Both flowers and leaves look similar. Note hairy stalk. Dandelion has a smooth stalk. Edible.
- **Sow thistle, annual** (*Sonchus oleraceus*) Both flowers and leaves look similar. Note spiky edges of leaves. Multiple flowers per stalk. Edible.
- **Sow thistle, prickly** (*Sonchus asper*) Both flowers and leaves look similar. Note spiky edges of leaves. Multiple flowers per stalk. Edible.
- **Hawkweed** (*Hieracium* spp.) Leaves are hairy and not jagged like dandelion. Multiple flowers rather than just one per stalk. Some species are edible.
- **Prickly lettuce** (*Lactuca serriola*) Leaves look similar when young. Note spines on underside of leaf. Edible when young. See prickly lettuce chapter on page 49.

Asparagus

Asparagus officinalis | Asparagaceae

Synonyms

Garden asparagus, sparrow grass

Description

Perennial (up to 2 m). Likes the edge between the forest and field in areas with good sunlight. **Shoots** are thick and fleshy. When snapped, they exude a lot of juice. Their color varies from green to purple. When the plant gets older, it fronds out and becomes tall, green, and bushy. Its **leaves**, which really are modified stems, are needle-like and give the plant the appearance of a fern. Plant is dioecious. Male plants tend to produce larger shoots. **Flowers** are small, white, and bell shaped. They have six tepals that are partially fused together at the base. Male flowers have six stamens with yellow anthers. Female flowers have a pistil with a single style. Female plants will have red **berries** later in the year. Inside the berries are 2 to 4 hard black **seeds.** The **root** is called a crown, which looks to me like a giant spider. Asparagus can also send out **rhizomes** and form clones of itself.

Pollinated by insects and is a favorite with bees. An asparagus plant can live and produce edible shoots for 20 years if well-tended. Young plants do not produce shoots for the first 2 to 3 years of their life. *Asparagus* comes from an ancient Greek name for the plant, and *officinalis* indicates its pharmacological use.

Uses

Today, the kids and I took our baskets and went hunting for asparagus in the field. We found enough to fill a basket. The trick to spotting them is to note the yellow stalk from last year that functions as a flag for these tasty spring treats. Once we find them, we take turns snapping them off. What a satisfying snap they make! The trick is to go down the stalk and let it snap

Figure 16.1. Mature asparagus plant.

Figure 16.2. Shoots.

Figure 16.3. Flowers attached in twos.

Figure 16.4. Berries on female asparagus plant.

where it wants to. If you try to force it to break further down the stalk, this will only result in an inedible fibrous part that is better left behind. Do not try to yank out asparagus as this will only damage the roots. You want that precious wild perennial to come back for many years to come. We harvest the shoots every few days for a period of three weeks or so. After that we let the plant frond out and get busy doing its business of creating flowers, pollen, and eventually seeds. If you find asparagus that has already furled out its side shoots, it will be too woody to consume. Best course of action is to remember the spot and return the following spring for a scrumptious harvest.

Later in the year, the female plants will have red berries. These berries remain on the fronds throughout the winter and can sometimes be seen the next spring. The berries are said to be poisonous, but the seeds can be roasted and made into a coffee substitute. It tastes surprisingly like coffee. I dry the berries and then rub them to remove the flesh of the berry. The seeds are very hard and shiny black. I put them in a frying pan and cook them like you would popcorn until they pop. At this point, they are brittle and can be crunched between the teeth. I grind them with the mortar and pestle and then brew like regular coffee. It is a very satisfying drink. Dark and wholesome with just a hint of bitter. I like to drink it with a splash of milk.

We make sure that we scatter some seeds around to ensure that the plant gets to reproduce. I like going out in autumn with the kids on a seed scatter mission. The only goal is to help the wild plants we harvest throughout the year to spread their seeds. It is a thank-you gesture to show how incredibly grateful we are to the

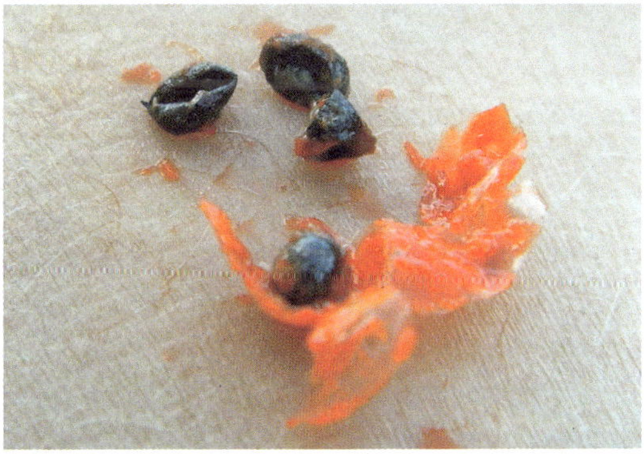

Figure 16.5. Seeds. Remove all berry flesh before using.

Figure 16.6. Roasted seeds can be ground and used as coffee.

plants. It is also a great exercise for the children to be able to spot different habitats and think about why certain plants like to grow where. I have been collecting some of the seeds from a variety of wild asparagus, and I am starting them as seedlings with the intention of putting them in my garden bed to have a ready-at-hand supply. The wild asparagus differs very little from the cultivated kind and can be used the same. The seeds take about two to three years before they start to produce shoots that you can eat, so it is a bit of an exercise in patience. Luckily for us, there are lots of wild ones out in the field to harvest until then.

Other uses

Asparagus is high in vitamins B6, E, and C, calcium, dietary fiber, and an amino acid called asparagine that is important in development and function of the brain. Asparagus has been used medicinally for over 2,000 years. It is used to cleanse the urinary tract. It helps to flush the body of excess salt and can regulate blood sugar. The shoots have antioxidants and anti-inflammatory properties.

Figure 16.7. Tasty asparagus shoots.

Asparagus and tomatoes are good companion plants, helping each other to limit pests. Asparagus contains asparagusic acid, which has nematocidal properties that deter nematodes or roundworms from eating the plants.

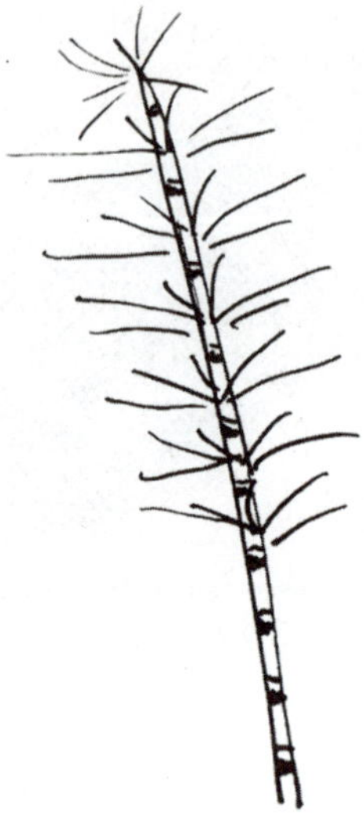

Figure 16.8. Horsetail (*Equisetum* spp.). Note the segmented way in which it grows.

Similar species / look-alikes

- **Horsetail** (*Equisetum* spp.) Grows much smaller and has segmented stalks that feel rough to the touch. Will not produce flowers or berries. Some species are edible with caution.
- **Canada fleabane** (*Conyza canadensis*) At first glance might look similar, but this plant has soft small leaves and composite flowers at the top. Does not produce berries. Edible.
- **White sweetclover** (*Melilotus alba*) The dried stalk early in spring looks similar and has at times tricked me into searching for the shoots at the base of the wrong plant. In their prime, they look nothing like one another. Edible. See white sweetclover chapter on page 117.

Caution

Do not be alarmed if your urine starts to smell a bit peculiar after you have eaten asparagus. This is completely harmless and disappears after a few hours. Certain chemicals in the plant, when digested, break down into a related group of sulfur compounds that cause the smell.

Be cautious with this food or avoid it if you have kidney stones. The red berries are mildly poisonous. Do not consume. They can cause vomiting and abdominal pain. If you use the seeds make sure to remove all the berry flesh.

I eat the asparagus shoots raw, but some people have an allergic reaction to them. Those who are sensitive to asparagus can also develop contact dermatitis when touching the plant.

Prickly Lettuce

Lactuca serriola | Asteraceae

Synonyms

Milk thistle, compass plant, scarole

Description

Tall (1 to 1.4 m) biennial plant that likes to grow in disturbed soil. When young, the **leaves** form a basal rosette. **Leaves** are blue green with spikes on the edges and barbs on the underside. Older plants have a pinkish spiky **stem** with leaves attached in an alternate pattern. Each leaf has long auricles at its base. Both stalk and leaf seep a milky sap when broken. Stem terminates in a panicle with many yellow composite **flowers**. The grey-brown **seeds/achenes** are attached to small white tufts of hair. The length of the seed is shorter than the threadlike stalk that attaches it to the hairs/pappus. It has a sturdy yellow **taproot**.

Lactuca means "milk," and *serriola* means "serrated margins."

Uses

I marvel at the abundance of prickly lettuce on the farm. Some of this lush and juicy wild lettuce has taken hold in one of my garden beds, and it does not look out of place next to my salad mix and mizuna. This species is one of the ancestors to cultivated lettuce (*Lactuca sativa*). The flavor is sometimes mild, sometimes bitter, and it is reminiscent of regular lettuce, save for the barbs on the underside of its leaves. No one in my family seem to object to the (soft) barbs, but if they bother you, add this plant to cooked dishes where you will not notice the barbs. It can be eaten when young but should not be used as food after it has flowered. At this stage, it can cause stomach upset and nausea. The plant is slightly sedative, meaning it makes you calm and drowsy. In order to make sure we do not overconsume this plant, we do not eat it every day. We might eat prickly lettuce twice a week during the early spring.

Figure 17.1. Young prickly lettuce leaves, perfect for salad.

Figure 17.2. Underside of older leaf: note barbs.

Figure 17.3. Milky sap/latex.

Figure 17.4. Seeds/achenes. Note the long stalk of the pappus.

Other uses

When the plant is flowering, the concentration of lactucarium in it increases. Also called lettuce opium, it has similar qualities as opium and works as a sedative, relaxant, and analgesic. It is not addictive but should nonetheless be treated with caution and respect. Too much of it can cause cardiac paralysis or restlessness. In small doses, it can help to alleviate pain. It acts on the central nervous system to calm the nerves that are experiencing the pain. I use this tincture in very small quantities. No more than three drops at any one time, and I do not use it more than three times a day or for more than three days in a row. I do not give it to my children. I have found that it alleviates back pain, but I have not used it

Caution

Be aware that this is a potent plant and should only be used in small quantities. It can cause lethargy, vivid dreams, balance loss, and distorted vision. It can also cause upset stomach. If too much is ingested, it can cause restlessness, cardiac paralysis, and death. Since I am neither a herbalist nor a doctor, I do not want to give you specific dosage as I do not know your weight, health, or any allergies. I can only tell you what works for me. Please do further research on your own. When used in moderation, this plant can be very helpful, but it should be treated with respect. Contains more lactucarium when in flower and should not be used for food then.

Figure 17.5. Prickly lettuce flower.

Figure 17.6. Stalk and leaf: Note the clasping manner of the leaf.

very much, so I cannot say for certain whether it is the tincture that is working or other variables unaccounted for. It has been standardized, meaning it is used in pharmaceuticals available in stores. It can be found in lozenges, tinctures, and syrups for cough, insomnia, and rheumatism, as well as in commercial preparations for pain relief.

Prickly Lettuce Tincture for Pain

- 2 fresh plants (minus the roots), chopped into smaller pieces
- Put enough pieces in a large jar to loosely fill it.
- Cover with vodka to the top to avoid air getting to the plant matter.
- Label the jar—important!
- Store in cool, dark place 6 to 8 weeks. Give it a shake every few days.
- Strain and use as needed.

Similar species / look-alikes

- **Bitter lettuce** (*Lactuca virosa*) Almost identical. Taller than prickly lettuce, can be 2 m tall. Seeds/achenes are purple black rather than grey brown. More common in Europe. Used similarly. Edible.
- **Wild lettuce** (*Lactuca canadensis*) Does not have barbs on underside of leaf. Achenes are attached with a much shorter stalk to pappus. Achenes are darker in color. Used similarly. Edible.
- **Dandelion** (*Taraxacum officinale*) Has only one flower per stalk. Stalk is hollow. No barbs on underside of leaf. Edible. See dandelion chapter on page 41.

Figure 17.7. Chicory, dandelion, prickly sow thistle, and prickly lettuce flower comparison. For leaf comparison, see page 44.

- **Chicory** (*Cichorium intybus*) Has blue flowers. Hairs on leaf midrib instead of barbs. Edible. See chicory chapter on page 163.
- **Sow thistle** (*Sonchus* spp.) Leaves have spikes on edges, not on midrib underside. Edible.

Black Elderberry

Sambucus racemosa var. *melanocarpa* | Adoxaceae

Synonym

Rocky Mountain elder

Description

Medium-sized tree/shrub. Grows in moist areas. Bipinnate **leaves** opposite on branch. Usually, it has five leaflets on each leaf, sometimes more. Leaves are elliptic with jagged edges. Top of the leaf is uniformly green; the bottom is paler and has fine fuzz. Tree/shrub has a musty scent. **Flower** is white with five petals. It has five yellow stamens jutting out between the petals and a white pistil in the middle. Elder is monoecious; both female and male organs on the same tree. Flower is tiny and carried in a pyramid panicle/compound raceme. Pollinated by insects. Pedicels are green during the flower period and sometimes turn pink/purple when the berries are formed. **Berries** start out small and green, and later develop into dark red/purple, sometimes black berries.

Elder belongs to the family Adoxaceae. They were formerly included in the Caprifoliaceae family. Both these families have opposite leaves and **pithy stems**. The difference between the two is that the Caprifoliaceae/honeysuckle family has flowers in pairs and the Adoxaceae family has flowers in clusters.

Sambucus comes from "sambuca," a stringed instrument, similar to a harp, that was typically made from elder wood. *Racemosa* means

Figure 18.1. Elder shrub.

Figure 18.2. Flower. Note how flowers are arranged in a pyramid panicle.

Figure 18.3. Leaf, pinnate with five leaflets.

"having racemes," while *melanocarpa* translates as "black or dark fruit."

Uses

Today while trying to get my toddler to fall asleep in the stroller, I wandered past the elder bush and was surprised at how early the flowers had been and gone this year. I thought I had plenty of time yet to gather this precious gift, but it looks like I will only grab the tail end of the flowers this year. This elder species (*Sambucus racemosa* var. *melanocarpa*) is earlier to flower and fruit than the other blue and black elders.

I like to use the flowers to make cordial. I place the flowers in a jar with water and let sit for a few days in the fridge. Then I add a little lemon juice and sugar; this is not necessary, but I think it tastes marvelous together. This is the tastiest drink you will ever come across. Nothing spells the beginning of summer as an elder cordial! When you harvest the flowers, make sure that you do not harvest all from any one tree; each flower you harvest is a loss of a berry later in the year. Not only do you want to be able to harvest this lovely gift as well, but it is also an important food for many of the forest critters.

The berries ripen in July/August, and they can be used to make jams, jellies, juice, and wine. They are low in pectin so are best mixed with other fruits if you are making jams and jellies. Their musty flavor takes a little bit getting used to. Once you have developed a taste for them, I think you will come to appreciate their more nuanced flavor. The berries are a good source of potassium, antioxidants, iron, and vitamins A and C. Some people experience mild nausea when eating them raw, so try only

Flower Fritters/Pancakes

1 cup elder flowers, removed from the stems
1 cup flour
1¼ to 1½ cups milk
1 egg
1 tablespoon oil
salt

- Mix flour, salt, egg, milk, and oil until you have a smooth pancake batter. Add seeds if you wish. I like using chia seeds. Gently mix in the elder flowers.
- Fry in mini pancake form. Lilac flowers (*Syringa vulgaris*) can be added to this as well. They both flower at the same time, and the color contrast is a delight.
- Serve with your choice of syrup or jam.

Figure 18.4.

Figure 18.5. Wild jam from wild raspberry, thimbleberry, elder, red-osier dogwood berries, and green apples.

a few initially to see how you react. I personally have never experienced any nausea or vomiting from ingesting the raw berries. The seeds become harmless once the berry has been cooked or dried, since the cyanide in them disappears almost entirely through the process. The dried berries make an excellent tea during winter months.

Other uses

The flowers are diaphoretic, which means that they increase body temperature, which aids in killing flu bugs. I mix these dried flowers with peppermint and yarrow for a great fall flu tea. From the dried berries, I make a tincture. All the dark, blue, or purple varieties can be used for this purpose. I add them to vodka or rum and leave for a few months to soak, stirring or shaking the jar occasionally. You can also make a tincture of fresh berries, but since there is cyanogenic glycoside and poisonous alkaloids in the stems and seeds, I feel it best to dry the berry clusters first. The bonus of drying the berries first is that they are much easier to remove from the small stems. Trying to do this while they are fresh is a tedious and messy project. Personally, I leave the berries in the tincture, but I know others strain them out and decant in different bottles with droppers. I just put my dropper in the jar, and whenever I feel a little run-down, tired, or the kids start to have runny noses, I take a dropper full of the tincture, and I feel that this keeps me from catching any colds. I also give my kids a drop each day when they start to develop a cold or when there are a lot of colds going around. I do not give it to them more than seven days in a row.

Elderberries have been tried in clinical tests and have been proven to combat up to eight different strains of flu. In a small-scale experiment, it was also shown to be helpful to combat H1N1.

Another lovely use of the elderberry tree is as a dye. The black berries make any color from dark purple to pink, depending on how much you crush them and how long you let them soak with the fabric/yarn you want to dye. The leaves can also be used and give a green/yellow color, and the bark produces a grey hue. According to the description I follow, it is sufficient to use vinegar as the mordant.

Elder branches can also be made into flutes. They have a pithy interior that is easy to remove, which gives you a hollow stem, perfect to craft into instruments. A word of warning: if you want to try this yourself, let the branch dry thoroughly before you put your lips to it. There is a slight chance of poisoning if you get the toxic from the fresh branch into your body. I tried to make a flute but was never able to get any sound out of it. This says more about my abilities than it does the elder's properties.

Since we raise sheep on the farm, I was interested to read that if the sheep have access to elder bark and young shoots, they will cure themselves of foot rot. Although foot rot has not been an issue thus far for us, I will keep this information in mind.

Figure 18.6. Early spring black elderberry buds.

Caution

The shrub contains cyanogenic glucosides and a poisonous alkaloid in leaves, stems, bark, roots, and seeds. The cyanogenic glycosides in the plant break down and release cyanide when it is injured. Some people experience nausea, vomiting, and diarrhea when ingesting too many berries. It is safer to dry or cook the berries before ingesting.

Figure 18.7. Black elderberry (*Sambucus racemosa* var. *melanocarpa*) berries with pink-purple pedicels/ stalks.

Similar species / look-alikes

- **Red elderberry** (*Sambucus racemosa* var. *pubens*) Has red berries. They form in racemes or pyramidical clusters just like *Sambucus racemosa* var. *melanocarpa*. Flowers can be used in the same manner, but seeds have higher concentration of cyanogenic glycosides. Berries can be juiced and used to marinate fatty fish with great results. Can cause nausea in some people. Use with caution.
- **Blue elderberry** (*Sambucus cerulea*) Blue, white-coated berries and white creamy flowers in large cymes. They have a flattened top rather than a pyramid shape. Can be used in same manner.
- **Common elderberry** (*Sambucus canadensis*) Dark purple to black berries. The white flower clusters (cymes) have a flattened top rather than a pyramid shape. Can be used in same manner.
- **Black elderberry** (*Sambucus nigra*) Introduced European species, often grown as an ornamental in North America. This elder can be told apart from the common elder by its blacker berries. They are borne in a cyme rather than a raceme. Can be used the same.
- **Common privet** (*Ligustrum vulgare*) Introduced species from Europe. Privet has tubular flowers with four petals that are bent backwards. Their smell is unpleasant. Leaves are small with entire margins. They are arranged in opposite pairs on the branch. Berries are dark black and borne in panicles. Inside are 1 to 4 seeds. Berries are **toxic** to humans. See photo on page 116.

Catnip

Nepeta cataria | Lamiaceae

Synonyms

Catmint, cat wort

Description

Perennial member of the mint family (up to 1 m). Has a square **stalk** and fine fuzz covering the entire plant. **Leaves** are triangular with jagged edges, very soft to the touch, and emit a strong minty/musky smell when bruised. Leaves are bluish green; they grow in an opposite pattern. **Flowers** are white and tubular with small purple/pink blotches on them. Popular with bees, bumblebees, and butterflies. **Seeds** are black/brown, small and hard. It has a **taproot**, but any stem can form new roots if layered on the ground. Can be invasive.

Nepeta is thought to come from Ivy of Nepete. Nepete was an Etruscan town. The Etruscan civilization was in ancient Italy. *Cataria* refers to cats.

Figure 19.1. Catnip — a favorite with cats.

Uses

We harvest catnip leaves all throughout the spring, summer, and fall. I like to gather my leaves from plants that have yet to flower because I find the plant softer and juicier then. We gathered enough leaves today to make a batch of catnip cookies. Use any cookie recipe that you have, be it butter cookies, shortbread, or oatmeal cookies. Any of these recipes will work. We used an oatmeal recipe and replaced half a cup of oats with one cup of fresh chopped catnip. The result was delicious. You can really taste the catnip in them. Kind of minty, with undertones of some leafy green vegetable. You could also use dried catnip, in which case, I would not use more than half a cup because the flavor could be really overpowering if too much is added. Try different recipes and see which one you like the most! Isn't that the beauty of cooking and

Figure 19.2. *Left to right*: Catnip leaf top and bottom. Square stalk.

baking? Finding novel ways to put tried-and-true ingredients together and then add a twist of the wild.

The young leaves are also a nice addition to a salad. A little too fuzzy to eat in great abundance but chopped up in small pieces and added to mixed greens, they provide an interesting, refreshing flavor. We also use small amounts in our cooking throughout the year. I toss both dried and fresh catnip into omelets, casseroles, soups, pancakes, and bread. The leaves go well in our lamb stew. I have not yet made catnip jelly, but it is worth a try, I think.

We also like to gather the flowers later when they are in full bloom and add to our salads.

Figure 19.3. Catnip leaf.

Other uses

Catnip has calming properties and works great as a relaxing infusion for the kids. An infusion is basically the same as a tea: pour hot water over the leaves and cover the cup so that none of the essential oils escape with the steam. Boiling catnip will destroy some of its beneficial properties. Catnip is soothing to the nervous system and can help alleviate anxiety and hyperactivity. The chemical in catnip that has the analgesic and sedative effect is thought to be nepetalactone. It functions as a slight analgesic, meaning that it takes away pain by blocking prostaglandins that constrict the blood vessels. When the blood vessels can open more and receive more blood to the part where the pain is located, less pain is felt. This could be why it is also indicated for headaches and swelling.

When you feel a flu coming on, with upset stomach, fever, and achy joints, then catnip is your buddy. It brings up your temperature which helps to fight off those bad germs. It can be mixed with elder flowers (*Sambucus* spp.) for added efficacy against the flu. It soothes

Caution

If ingested in large doses, catnip can cause nausea. There are different opinions about using this herb during pregnancy. Some sources think it safe; others warn against it. I avoided it while pregnant.

Figure 19.4. Flower.

digestive upsets and can help with nausea, flatulence, and diarrhea. I avoided it while I was pregnant though; any mint gave me an uncomfortable feeling.

The essential oil from this plant can be used to repel mosquitoes. It is also reputed to deter ants, fleas, cockroaches, and mice. I am not sure how effective it is. I have seen mice scurrying around among the catnip, so perhaps the real deterrent are the cats the plant attracts. Some cats experience euphoria when exposed to the plant, others have no reaction at all. This depends on specific genes in the individual cat.

Similar species / look-alikes

- **Lemon balm** (*Melissa officinalis*) Garden herb that looks very similar. Leaves are brighter green, and it lacks the fine fuzz that catnip has. Lemony scent. Edible.
- **Mint** (*Mentha* spp.) All mints are very similar looking. All of them have the square stem, and most have the telltale minty smell. They also crossbreed, so an extensive description of all is beyond the scope of this chapter. Most are edible, but some are too potent.
- **Stinging nettle** (*Urtica dioica*) Stinging nettle leaves are dark green, catnip leaves are more blue-toned. The most obvious way to tell them apart is that stinging nettle has stinging hairs, and catnip is covered in benign velvety fuzz. Nettle also lacks the minty scent. Edible. See stinging nettle chapter on page 36.

Red Clover

Trifolium pratense | Fabaceae

Synonym

Purple clover

Description

Introduced perennial plant (up to 60 cm tall). Prefers sunny exposed areas and disturbed soil. Has three oval **leaflets** with white crescent markings. If you are lucky, you might just find one that has four or more leaflets. At each node, two stipules clasp the stem. Both stem and leaves are coated in fine hairs. At the top of the stems are round balls that consist of many small **flowers**. These are red, pink, or purple. Each individual flower has five distinct petals: a banner petal on top with red lines running down it, two side petals in a lighter color, and a bottom petal called a keel. The bottom of the flower is tubular and holds the delicious nectar. This plant is an attractant for numerous pollinators. **Seed** is small and roundish. Color can range from black to cream. **Stalk** is yellow green and stringy. **Root** is a taproot with supporting roots. Attached to these are small round **nitrogen-fixing nodules**. When the plant decays, that nitrogen is released into the soil and becomes available to other plants. Hence, it can be beneficial as a green manure.

Trifolium means "three-leaved." *Pratense* means "of the meadow."

Uses

Both the bumblebees and the kids are excited about this sweet treat. They swarm the plants together, picking off the petals and sucking out the sweet nectar from the bottom. We collect the flower petals and use them raw in salads and as decoration on ice cream and cake. They lose

Figure 20.1. Red clover flowers.

Figure 20.2. Two stipules clasp the stem at each node.

Figure 20.3. Three leaflets with crescent markings.

their extraordinary color when exposed to heat so are better used raw. I dry the flower heads and use them for tea throughout the year. They taste sweet in a mild way.

Another lovely use of the entire flower head is to make syrup. Below is a recipe that makes a delicious syrup (a little too sweet for my liking, but great on pancakes). It reminds me of sticky honey in texture.

We also use the leaves for food, and these we pick before the flower emerges. Once the flower appears, the raw leaves are not as easy to chew; they become more fibrous. The flavor remains

Caution

An important note on using clover (this applies to all the clovers) is that great care must be taken to ensure that there never is any mold in the leaves or flowers while drying. If mold is present, the coumarin (which is what the vanilla scent comes from) is converted into dicoumarol, which is an anticoagulant; it can hinder blood clotting and result in excessive bleeding. Red clover can interact with prescription blood thinners.

Ingesting too much raw clover leaves can cause bloating, but this does not seem to be a factor if you cook the leaves first.

The seeds are best sprouted before being used since they contain trypsin inhibitors that can interfere with the enzymes that aid in the absorption of protein. When the seeds are sprouted, this is not an issue.

Clover Flower Syrup

2 cups water
2 cups flower heads (all green leaves removed)
4 cups sugar
lemon juice

- Cook water and flower heads in a pot on low for 20 minutes. All the color drains into the liquid, and the flower heads turn grey.
- Discard flower heads and keep the vibrant orange-red liquid.
- Cook on stove top with sugar (white sugar allows the flavor to come through the most).
- Add a splash of lemon juice. Boil 10 minutes. Can or freeze.

the same though, and they are still usable, especially in dried and powdered form. Raw clover leaves taste green and healthy, what you would imagine a meadow to taste like. I use the leaves in cooking, both fresh and dried. They can be turned into a protein-rich powder that can be added to breads, muffins, smoothies, and soups. I think that the dried leaves are tastier than the fresh ones. They have an added sweetness and smell lovely of vanilla due to the coumarin in them. See caution box.

The seeds of red clover can be collected and used to sprout as you would alfalfa seeds. **Sprout before use—see caution box.** I find them to be too fiddly for a successful harvest. I had too much chaff in the ones I collected, and the chaff molded before I could get the seeds to sprout.

Figure 20.4. Clover flower syrup.

Figure 20.5. Dried red clover flowers. Store in dark place to retain color.

Other uses

Red clover has been used extensively to regulate estrogen and ease menopause symptoms. I have also read that it is supposed to help with conditions such as eczema and psoriasis. We are just starting to use it for eczema in our family, but I do not have conclusive results yet to attest per-

Figure 20.6. White clover.

Figure 20.7. Alsike clover.

Cough Syrup/Honey

3 cups water
1 cup dried clover flowers
1 cup dried whole mullein leaves (*Verbascum thapsus*)
1 large dried thyme sprig (*Thymus vulgaris*)

- Combine and simmer on low heat for approximately 15 minutes.
- Strain the liquid through a coffee filter (about 2 cups remaining) to remove any hairs from the mullein leaves, as they can irritate the throat.
- Pour the hot liquid over 2 cups of honey and dissolve gently. Stir as it cools to stop it from separating.
- Store in jars in fridge and use when you want a respiratory aid or take a spoonful when you feel a scratchy throat coming on.
- If you want a more potent syrup/honey with added medicinal qualities, add 2 tablespoons of dried horehound (*Marrubium vulgare*) to the mixture. It will not taste as good, but it will be even more effective at clearing the airways.

sonally to its efficacy. Red clover contains vitamins A, C, F, P, and B-complex, as well as iron, magnesium, calcium, and copper.

Similar species / look-alikes

- **White clover** (*Trifolium repens*) Grows low with creeping stems. Flowers are white rather than pink and are much smaller. Leaves have the distinct three-leaf clover shape with white markings. Leaves are rounded and dark green. Edible.
- **Alsike clover** (*Trifolium hybridum*) Flower is white with pink at the base. Does not have the white markings on the leaflets. This plant is a hybrid between red and white clover. Edible.
- **Alfalfa** (*Medicago sativa*) Much taller than clover, and the flower clusters are not in tight balls like the red clover. Flowers are blue purple. Leaflets are skinnier and toothed at the tip. Seed structure is a spirally coiled pod. Edible.

Figure 20.8. Alfalfa.

Salsify

Tragopogon dubius | Asteraceae

Synonyms

Yellow salsify, western salsify, yellow goat's beard, wild oyster plant, sand goat's beard, fistulous goat's beard

Description

Introduced biennial herb (up to 1.2 m tall). Likes to grow in sunny spots. **Flower** is yellow and looks like a mixture between a dandelion and a daisy. Has a row of larger yellow ray florets and numerous rows of smaller yellow ray florets. There are dark brown stamens in the center of the flower. Sepals are green and longer than the petals. They stick out from behind the yellow petals giving the plant a star-like look. Flower opens when sunny and closes early in the day. When closed they look like green seedpods. **Stalk** is hollow, thick and blue green. Plant oozes white milky latex when cut. **Leaves** are skinny with straight venation, and they clasp the stems; they remind me of grass blades. **Root** is a fleshy taproot. Brown **seeds** are attached to long-stalked pappus. Seed balls shimmer in gold. Interestingly, salsify creates two kinds of seeds in the seed ball. In the middle are lighter and less textured seeds, these weigh less and can travel further on the wind. On the lower part are larger, darker, and heavier seeds. These tend to end up closer to the parent plant. Although salsify is technically a biennial, some reports say that it can take more than two years for a plant to flower. Some have even taken fourteen years before flowering. Once it flowers and sets seed the plant wilts and dies.

Figure 21.1. Note salsify's long sepals.

The name salsify is formed from an old Latin name *Solsequium*. *Sol* means "sun," and *sequens* means "following." The name *Tragopogon dubius* comes from the Greek *tragos* ("goat") and *pogon* ("beard"). The *dubius* part means "uncharacteristic" or "dubious."

Figure 21.2. Leaf with straight venation.

Figure 21.3. Digging for salsify root beginning of June.

Uses

We use all parts of the salsify. Make sure you collect the root before the plant has flowered. If you try to harvest it when the plant blooms, you will get a spindly little woody root. Without the flowers, the plant can be hard to detect in a grassy field. It disguises itself as one of the grasses. Once you have spent some time with salsify though, you will be able to spot it quite readily; it is bluer than a lot of grasses. Its taproot can at times go deep down, and in hard-packed dry soil, it can be particularly hard to yank out. Be mindful that when you harvest a root from a plant, you are killing that plant. To ensure that you are not depleting the species, only harvest a few plants in each stand. We also gather seeds and scatter them to help the plant reproduce as a thank-you for its scrumptious root. It is fun to see the seeds gather height on the slightest upwind.

To prepare the root, I gently scrub it and then cut it finely. It tastes wonderful oven-baked with other root vegetables and coconut oil. Tomato and salsify root soup is a nice combination that is both hearty and light. With some shaved cheese and a fluffy baguette on the side, it is a decadent meal.

I like eating the leaves during the spring and summer. They are mild flavored and can be eaten both raw and cooked. The flowers can also be eaten, in all the ways that you use dandelion. I like coating them in flour and frying them in butter. They also taste nice added to an omelet or as a crêpe filling. Even the stalk is edible; I use the top portion with the flower. Further down, the stalk can be a bit too fibrous. I cook the stalk and connected flower briefly in some water and then toss with butter, parmesan cheese, and a sprinkle of lemon juice. I serve it as a tasty and interesting-looking side dish.

Figure 21.4. Roots.

Figure 21.5. "Wild scrambled eggs" with salsify leaves, oxeye daisy, thistle, dock, shepherd's purse, chickweed, and red root pigweed. Lilac flowers for edible garnish.

It is thought that early settlers brought salsify to North America as a food plant. During the Middle Ages, it was commonly eaten in Northern Europe. The plant is originally native to Eurasia and Northern Africa.

Other uses

Salsify has cleansing qualities, and the root has been used in the past to aid the liver and gallbladder in their functions. Some sources say that it can be used for heartburn and indigestion. The root is a diuretic, i.e., it increases urination and has been used to help in treatment of kidney stones.

Grouse, pocket gophers, deer, squirrel, and rabbits are some of the animals that consume the plant.

Figure 21.6. Cut salsify leaf, note the milky latex.

Figure 21.7. Seed ball in July.

Similar species / look-alikes

- **Meadow salsify** (*Tragopogon pratensis*) Very similar, but the green sepals do not extend past flower head. Flower stalk is not hollow. This plant sometimes hybridizes with *T. dubius* and will then show characteristics between the two parents. The hybrid's name is *Tragopogon miscellus*. Both are edible.
- **Common salsify/oysterplant** (*Tragopogon porrifolius*) Flower differs in color; it is purple rather than yellow. This plant can hybridize with *T. dubius* and is then classified as *Tragopogon mirus* and will show characteristics between the two parent plants. Both *T. mirus* and *T. miscellus* have successfully reproduced in the wild, so they are considered species in their own right. Edible.
- **Dandelion** (*Taraxacum officinale*) Leaves are jagged and form a basal rosette. Hollow stalk with only one flower on each stalk. Seed head is similar but much smaller. Edible. See dandelion chapter on page 41.

Lamb's Quarters

Chenopodium album | Amaranthaceae

Synonyms

Fat hen, goosefoot, pigweed, manure weed, melde

Description

Annual herb that prefers disturbed soil (1 to 1.5 m tall). The blue-green **leaves** are rhomboid/diamond shaped with wavy edges and grow in an alternate pattern. They are smooth, matte, and coated with a powdery substance. When small, they are purple on the underside. When plant is older, the lower leaves are rhomboid, and the leaves on the stalk taper off and become lanceolate. Its green **flowers** are minuscule and

Figure 22.1. Note the white mealy powder on lamb's quarters leaves.

grow out of green clusters called glomerules that form in spikes at the top of the plant. Flowers have five sepals and no petals. They produce small yellow balls of pollen. Plant is wind pollinated and produces copious amounts of **seeds** that are small and black, shiny, round, and hard. Green **stalk** is fibrous and smooth, often with purple-pink stripes running lengthwise. It has a **taproot** and lots of supporting roots.

Chenopodium in Greek is made up of two parts: *chen* meaning "goose" and *podium* meaning "little foot," and alludes to the leaves resembling goose feet. *Album* means "white."

Figure 22.2. Green glomerules with flowers.

Uses

Without lamb's quarters, my existence would be much bleaker. I know many gardeners swear about it because it can be quite invasive, and it grows quickly, much quicker than those poor little carrots you are trying to cultivate. In the true spirit of a weed/pioneer species, it can create thousands upon thousands (500,000) of little seeds from each plant. They can remain viable in the soil for up to forty years and can survive intact through many kinds of digestive systems. So, it is not without reason that this plant can make gardeners shudder. However, when you start to get to know this plant better, you might be thrilled that it has taken up residence in your garden.

Figure 22.3. Young lamb's quarters leaves with purple tint.

This time of year, we eat the young seedlings both raw and cooked. They can be used in any way that you would use spinach. We also dry the leaves for winter use. I add them, crumbled up, to just about anything. Yesterday we had pizza with fresh lamb's quarters leaves and chickweed (*Stellaria media*). It was delicious. The flavor of lamb's quarters is very mild, so it is easy to incorporate into any meal. Some people find the powdery substance on the fresh leaves bothersome. That texture disappears when cooked. The leaves can be eaten raw when young. Once the plant flowers, the raw leaves sometimes

Figure 22.4. Shiny seeds.

cause a sharp scratchy feeling in the throat. They can still be used cooked or dried. Another way to use the dried powdered leaves is for seasoning. Lamb's quarters have mineral salts and taste great in an herb mixture to be sprinkled on food. The ashes of this plant can also be used as a salt substitute.

It likes to grow in disturbed soil, so your garden is a perfect habitat for it. The great thing about lamb's quarters is that it grows quickly and can become a good size for a meal fast. Because of its reliability and abundance (when my vegetables are still just wee ones), I can provide food for our family much earlier in the year than I would have without lamb's quarters. They are filled with nutrients and minerals. They contain calcium, phosphorus, iron, and vitamins A, B-complex, and C. They also contain a fair amount of protein.

Later in the year, we collect the young inflorescence glomerules (flower clusters before they flower) and use them in cooking. These we do not consume raw to avoid ingesting too many saponins. See caution box. The seeds are also a cherished part of our diet. To collect the seeds, I cut off the entire plant and put it headfirst into a paper bag. I let it dry some more in the paper bag and then rub to loosen all the seeds and husks into the bag. The only parts I remove are the larger stalks as they are hard to digest; the

Caution

Lamb's quarters contain oxalic acid. According to some sources, this can hinder absorption of calcium in the body. People with kidney stones may want to limit their intake of foods with oxalic acid because it can increase the production of kidney stones. Some of the oxalic acid is removed by cooking. Most people have no reason to worry about eating lamb's quarters. However, I would be more careful giving it to a very young child (4 to 8 months old) because of blue baby syndrome. This happens when there are too many nitrates in the baby's bloodstream. If you are not entirely sure of how much nitrate is in the soil where your lamb's quarters or spinach grows, I advise against feeding your baby these plants. Soils with added chemical fertilizer can be high in nitrates.

Lamb's quarters seeds contain saponins. Saponins in high doses can be irritating to the stomach lining. To avoid ingesting too much, soak the seeds for a day before using them and then discard the water.

rest is stored and used together (seeds, husks, inflorescence, and leaves). This mixture is very nutritious and great tasting in bread. I soak it overnight to remove saponins from the seeds. I then rinse and add to my bread dough. It tastes fabulous!

The seeds can also be sprouted during the winter and eaten as microgreens or as sprouted seeds in soups, breads, and smoothies. They do not take long to come up and require very little care to thrive.

Other uses

Some plants that contain saponins can be used instead of soap. I have seen reports of crushed lamb's quarters roots used for this purpose. Personally, I have never had any luck with the root. It is too woody, and I do not find any frothy substance coming from it. The seeds, however, make a lovely frothy hand cleaner. My hands are often stained from lack of using garden gloves while working. The seeds, and the water they have been soaking in, clean the hands and give them a nice scrub/exfoliation, leaving them feeling soft and squeaky clean.

Similar species / look-alikes

- **Redroot pigweed** (*Amaranthus retroflexus*) Often has a red-tinged stalk. Leaves have more prominent veins and lack the mealy powder. Seeds and flowers form in spikes with spine-tipped bracts that make them look bristly. Edible.
- **Oak-leaved goosefoot** (*Chenopodium salinum*) Grows lower to the ground in a sprawling manner. Leaves resemble oak leaves. Edible.
- **Strawberry blite** (*Chenopodium capitatum*) Has juicy red fruit at the top of plant later in summer. Leaves are heart or lance shaped. Edible.
- **Garden orach** (*Atriplex hortensis*) Leaves are heart shaped. Stem is purplish. It carries two types of flowers and two types of seeds. Edible.
- **Maple-leaved goosefoot** (*Chenopodium simplex*) The flowers and glomerules are not tightly clumped against the stalk but rather spread out. Leaves are shiny and bright green rather than matte and blue green. Can be used in same way.
- **Berlandier's goosefoot** (*Chenopodium berlandieri*) Identical in almost all ways apart from their seeds. This species has a seed that has a pitted surface rather than a smooth shiny surface. Very hard to see without magnification, and the two species also hybridize. Can be used in the same way.

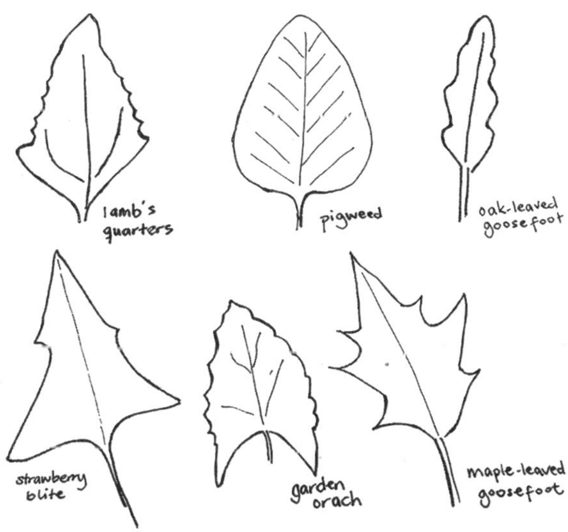

Figure 22.5. Comparison of leaves top to bottom: Lamb's quarters, pigweed, oak-leaved goosefoot, strawberry blite, garden orach, maple-leaved goosefoot.

Common Mallow

Malva neglecta | Malvaceae

Synonyms

Dwarf mallow, cheeseweed, cheeseplant, buttonweed, roundleaf mallow

Description

Annual or biennial (up to 60 cm). Introduced species from Europe. Can be quite invasive. Grows from a fleshy white **taproot**. **Leaf** is dark green at maturity, cup shaped/kidney shaped, with an opening at the bottom. Leaf has rounded lobes with a crenate edge (rounded teeth). Bottom of leaf has fine transparent hairs. Top is quite textured, like fine sandpaper. Flowers and seedpods are formed in clusters of six or more, on stalks in the leaf nodes. **Flower** is most commonly white but can also have a pinkish hue. Some have violet stripes. Flower has five white satin-like petals and five green sepals.

Figure 23.1. Top of common mallow leaf.

Petals are at least twice the length of the sepals. Anthers are white, and the stigma is pink. After the flower has been fertilized by bees, flies, or self-pollination, it forms the **seedpods/mallow cheeses**. They are green/yellow when immature and brown when ripe. They remind me of a sea anemone after its spines have fallen off. They sit protected in the sepals/calyx until ripe and ready to fall out. Once the seedpod has fallen off, it divides into small wedges (mericarps), which are the individual seeds with an encasing around them. The encasing is brown and slightly fuzzy, and the **seed** is dull black and round.

Malva refers to the Mallow genus, and *neglecta* means "overlooked."

Figure 23.2. Underside of leaf.

Figure 23.3. Flower.

Uses

We rely on common mallow for many meals. All parts of it are edible. The leaves grow in great profusion, which is lucky because when cooked, they shrink to almost nothing. The leaves have a mucilaginous texture and help to thicken any dish they are added to. They form the greater portion of our wild lasagna.

Another fabulous way to cook the mallow leaves is to lightly coat them in oil and salt, and then either fry them on high temperature or briefly roast them in the oven. It only takes a couple of minutes, and the result is a delicious chip that even the pickiest of eaters will devour. The leaves contain vitamins A, C, calcium, potassium, iron, selenium, and magnesium.

Dried leaves make a nice addition to bread, soup, veggie patties, and casseroles. Their taste is mild, so they combine well with more flavorful greens and vegetables. In a salad, the fresh leaves tend to be a bit too droopy for my liking. They do not have the crunch that miner's lettuce

(*Claytonia perfoliata*) or chickweed (*Stellaria media*) does. However, I do like adding a few of the very young leaves (1 to 2 cm across) to a salad. I add them whole, and they look like adorable little cups in the salad.

In a fresh salad, the flowers and immature seedpods shine. The seedpods look like small cheese wheels and are sometimes referred to as mallow cheeses. They do not taste like cheese, but rather like a crunchy pea. They can be eaten raw as long as they are green. We add them to yogurt and salad and throw them in pasta sauce and stir-fries just before serving. They can also be pickled like capers and used in the same manner. I most often take away the husk/calyx that they are contained within, but you can eat this part too if you find it too fiddly to remove.

The ripe seed can be ground up and used for food, added to bread or cereal. The flowers are delicate, sweet tasting, and look lovely in a salad or on top of chocolate ice cream. They can also be candied and used as garnish on cakes. They become mushy if they are heated up, so they are better fresh.

Several field guides mention that the root can be used instead of egg white to make meringue. To use it this way, you are supposed to boil the root in water and then strain the root out and use the water as an egg white replacement. When I have tried, it thickens somewhat

Mallow Lasagna

- Pick a lot of mallow leaves, more than you think you will need. The flowers and immature seedpods can be added too, or you can use them for other purposes. I get rid of most stalks because they can be a bit stringy, but some of the smaller stalks from the top of the plant can be cooked with the leaves and added to the lasagna.
- Cook your greens briefly with a little bit of oil. Set aside.
- If you feel like cheating, use a can of spaghetti sauce for your tomato sauce. Or make your own sauce. I use lasagna sheets that need no precooking for this recipe.
- Layer the greens with feta cheese mixed with grated cheese, pasta, and tomato sauce until you reach the top of your pan. The last layer is sauce and then cheese. Make sure that all the pasta is covered, if not, it gets hard and chewy.
- Cook for 40 minutes or so at 350°F (175°C), or until pasta is soft and cheese top has a nice golden look.
- I often add other greens to my mallow mixture to keep it interesting. Other plants that go particularly well in this dish are: plantain leaves (*Plantago major*), dandelion greens (*Taraxacum officinale*), clover leaves (*Trifolium* spp.), violet leaves (*Viola* spp.), thimbleberry leaves (*Rubus parviflorus*), dried or fresh stinging nettle leaves (*Urtica dioica*), young paper birch leaves (*Betula papyrifera*), and lamb's quarters leaves (*Chenopodium album*), to name just a few. Your imagination is the limit!

but does not become fluffy like meringue. The mixture works well in baking though. I made coconut tops using only mallow root decoction, sugar, pinch of salt, coconut flour, grated coconut, and coconut oil. Mixed and baked in the oven for 10 minutes, this gave us chewy little treats that went well with tea. I have also heard people using the seeds in the same way, but I have yet to try this. The original marshmallow was made from—you guessed it—marshmallow (*Althaea officinalis*), a family member of common mallow.

The root when cooked is quite fibrous, but the outer layer can be scraped off with the teeth, and this tastes to me like a mixture between corn and parsnip. If this seems too fiddly for you, the root can be thrown into your stockpot where it will impart a sweet flavor.

Other uses

Due to their mucilaginous properties, the raw or dried leaves can be used to soothe an infected throat. When made into an infusion, they coat the mucus membranes nicely and help to soothe a cough. The root can be used for this purpose too. I dry chunks of the root, and when someone has a scratchy throat, I cook the roots in water for 20 minutes, then add honey. This drink feels soothing for a raw, infected throat.

The plant can also be used for an upset digestive system, and it works as a mild laxative. I find this works well for the children. I throw both dried and fresh leaves into soup or bread or make tea and let the food be the medicine. The leaves have been used as a poultice, i.e., mashed and laid upon sores and swollen areas. It can be added to salves to soothe the skin.

Figure 23.4. Wild lasagna with mallow, violet, and plantain leaf.

Figure 23.5. Mallow chips, ponderosa pine crackers, goat cheese, and mallow cheeses.

Figure 23.6. Immature seedpods.

Figure 23.7. Roots.

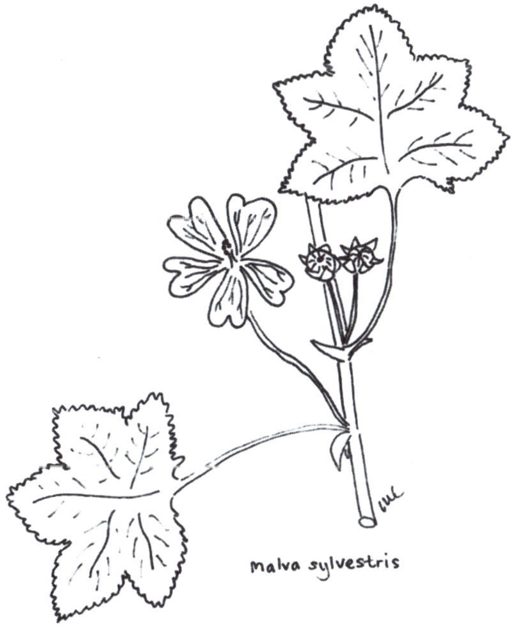

Figure 23.8. *Malva sylvestris*.

Mallow extracts are used in treating acne. The plant has antimicrobial and antifungal properties. The seeds have been shown to contain natural antioxidants and have been taken as a powder to help treat bladder ulcers. These last uses of mallow I have not tried myself.

Caution

When grown on nitrogen-rich soil, it can accumulate nitrates.

Similar species / look-alikes

- **Low mallow** (*Malva pusilla*) Looks almost identical. Petals and sepals are the same length (on *M. neglecta*, the petals are at least twice the length as the sepals). Flowers are smaller than *M. neglecta*. Seed encasing (mericarp) is rough and reticulated rather than smooth as in *M. neglecta*. To see this, you might need a magnifying glass. Can be used in the same manner.
- **Tall mallow/common mallow** (*Malva sylvestris*) Flowers are bright mauve/purple with dark red/purple streaks. It grows taller, up to 1.2 m. Leaves are maple leaf shaped with five distinct lobes. Can be used in same manner.

St. John's Wort

Hypericum perforatum | Hypericaceae

Synonyms

Perforate St. John's wort, common St. John's wort

Description

Perennial (about 60 cm). Introduced from Eurasia. Can be quite invasive in open fields with sun exposure and forms large stands. **Root** is a creeping **rhizome**. **Leaves** (1 to 2 cm) are oval with a smooth edge. If you hold them up to the light, you can see small dots/glands covering the entire leaf. Leaves grow in an opposite decussate manner (each set of leaves are at 90 degrees to the next set of leaves down the stalk), and they lack petioles.

Figure 24.1. Summer meadow.

Stalk is green, sometimes purple and woody. **Flowers** form during midsummer in broad clusters at the top of plant. One plant can have many flowers at various stages of maturity. Flower is bright yellow, shiny, and sticky to the touch. It has five petals and five sepals and numerous yellow stamens. The yellow pistil has three styles. Plant is monoecious (has both sexes on same plant) and is self-fertile. It is also pollinated by bees and flies. When the flower has been fertilized, it forms reddish husks and then seeds. Each plant can produce a large amount of seeds. **Seeds** are small, black, and pill shaped. When crushed, they exude a red oil.

St. John's wort was named so because the flower bloomed near, or on, biblical St. John the Baptist's day, the 24th of June. *Hypericum* comes from the Greek *hyper* ("above") and *eikon* ("icon"). The plant was hung over religious icons to ward off evil spirits. The dots/glands on the leaves are what gave the plant the Latin name *perforatum*, meaning "perforated" or "punctured."

Uses

At the end of June, St. John's wort is perfect for picking. We pick the top third of the plant and make sure that we leave stands to go to seed. I also use the dried leaves and flowers in cooking, adding them to frittatas and casseroles. I only use small quantities, and not very often, since over-consumption of this plant can cause side effects. See caution box.

St. John's wort has a pleasant flavor: sweet and astringent with a minor explosion of perfume inside your mouth. The fresh leaves and flowers can be added to a salad as well. The fresh flowers when added to vodka turn the alcohol red. Voilà, Christmas schnapps!

A cup of tea made from the dried twigs, leaves, and flowers seems to lift the mood somewhat. It tastes especially great with honey. I use it on a day when I feel like the darkness is closing in and I am having trouble seeing the bright side of life. I do not drink it for more than four days in a row, at a maximum of two cups per day. These are not prescribed dosages, just my personal usage. Sometimes it is enough to have a single cup of tea and hum Monty Python's "Always Look on the Bright Side of Life" to get life back on track.

Other uses

I also infuse the fresh leaves and flowers in cold-pressed virgin olive oil. After allowing them to dry overnight, I put them in a jar, pour the oil over them, and let sit for 2 to 3 weeks. I then strain the oil through cheesecloth. This oil can be used on its own, or it can be added to salves for skin care and to soothe itchy insect bites, scrapes and eczema breakouts. In the past,

Figure 24.2. St. John's wort flower.

Figure 24.3. Note the glandular dots on St. John's wort leaves.

Figure 24.4. Seedpods.

Figure 24.5. St. John's wort tincture after steeping for one hour. It will get dark red with time.

I have used St. John's wort oil (without adding it to a salve) as a night moisturizer.

This year I am going to try a new recipe called Green Ointment from Julian Scott's *Natural Medicine for Children* that calls for St. John's wort oil and elder leaf oil (*Sambucus* spp.) mixed with beeswax. It is supposed to be particularly good for itchy insect bites. My younger child has severe reactions to mosquito bites, so I hope this ointment will help her to not scratch the bites so much.

St. John's wort has a long history of use as a medicine, and many clinical trials have been done to test its efficacy. It has proven to be quite effective in treating patients with mild to moderate depression. See caution box. However, it should not be used with other prescription medication since it can interfere with or intensify that medication.

Caution

Ingestion can cause photosensitivity in some individuals. Excessive use over an extended period can cause fatigue, irritability, diarrhea, dizziness, and headache.

It should not be used with other medication, especially other medication intended for depression, since it can increase serotonin levels to unhealthy amounts. It can also interfere with other medications, rendering them less effective. Always consult with your doctor before self-medicating with this plant!

Figure 24.6. Sulphur cinquefoil, *left*; St. John's wort, *right*.

Similar species / look-alikes

- **Evening primrose** (*Oenothera biennis*) Flowers have four petals instead of five. Leaves lack glandular dots. Edible.
- **Sulphur cinquefoil** (*Potentilla recta*) Flowers are matte pale yellow rather than glistening bright yellow. Leaves are palmate. Does not have glandular dots. Edible. See sulphur cinquefoil chapter on page 257.
- **Buttercup** (*Ranunculus* spp.) Leaves are divided in three leaflets that are deeply toothed. It lacks the glandular dots of *Hypericum perforatum*. Used medicinally. Some species can be **poisonous**.

Figure 24.7. Buttercup.

Saskatoon

Amelanchier alnifolia | Rosaceae

Synonyms

Saskatoon serviceberry, serviceberry, shadbush, juneberry, chuckley pear, alder-leaved shadbush, pigeonberry, sugarplum, sarvis, wild-plum

Description

Shrub to medium-sized tree. Likes sunny spots. Showy white **flowers** with five petals early in spring. Petals are spread apart, and the green triangular sepals are clearly visible between each petal, giving the middle of the flower a star shape. There is one yellow/green pistil surrounded by numerous stamens with white filaments and brown/red anthers at maturity. **Leaves** are rounded with a blunt tip, dark green on top and whitish below. They have jagged edges and no hairs. **Bark** on branches is grey with a reddish tint, and the bark on the **trunk** is grey. Bark has lenticels/small slits that are white/yellow. **Berries** (pomes) turn from green to white to pink to dark purple/blue. They resemble small apples. Inside the berry there are 5 to 10 shiny, brown, wedge-shaped **seeds.**

Figure 25.1. Saskatoon flower in May.

Alnifolia means "alder leaf" (*Alnus* spp.), to which it bears a strong resemblance. *Amelanchier* is thought to be New Latin, meaning "shadbush, shadberry or serviceberry."

Uses

As the Okanagan heat kicks in, berries are ripening fast. The saskatoons are just turning from pink to purple/blue now. The kids and I were out eating them in the field today. It is still too early to harvest any great quantities, but somehow the first picking of a berry from bush to mouth is the sweetest of all. It is so important to also let the foraging be about being in the moment and enjoying the bounty without constantly rushing about trying to squirrel away as much as possible for the winter. Nothing is as sweet as lingering and eating ripe berries in the sunlight without a care for time or future. It is important when we pick them to be mindful that other species depend on those very berries for their survival, so that we never take more than our share. Some years the bushes produce more than others. On those years, it is ok to store away great quantities, but when the berries are scarce, just a handful here and there is enough for us.

The berries can be used in any way that you would use blueberries. Their seeds are larger and slightly harder than blueberry seeds. I like the extra crunch, and I feel they add some interest to jams and baking, giving cookies an almost nutlike flavor.

The First Nations Peoples of this area (Syilx and Secwépemc) have a long tradition of using saskatoon berries, one of their most important fruits. It features in the Captikwł stories of the Syilx. The Captikwł are creation stories that teach people how to live harmoniously on the land. Saskatoon (Siya) plays an important part in the story about the Four Food Chiefs and how people first found food. The berries were once an important trading item for both the Secwépemc and the Syilx People.

Before the berries appear, I sometimes harvest the young spring leaves, dry and then crumble them, and use in baking. They have a slight bitter almond taste due to their cyanogenic glycoside content. Cyanide is destroyed when cooking or drying, so I feel safe using them in small quantities on a sporadic basis. The seeds also contain a certain amount of cyanogenic glycosides. There is a debate whether small quantities of cyanide can be beneficial to your respiratory system, but I do not have enough knowledge to weigh in on that debate. I know that I have not suffered any bad effects from eating the seeds in the berry or from occasionally

Figure 25.2. Bud in February.

Figure 25.3. Frozen treat.

Figure 25.4. Saskatoon berries at all stages of maturity.

using the leaves in baking. If you are concerned, cooking the seeds and leaves remove most of the cyanide.

Other uses

The berries are rich in antioxidants and contain phenolics, flavonoids, and anthocyanins. These have beneficial effects for the body as anti-inflammatories, and they help to combat oxidative stress, in other words, help the body deal with detoxification and repair. Saskatoons also contain calcium, iron, and vitamin C.

The leaves can be used as an astringent to firm up tissue and reduce excess mucus in the respiratory system, according to Beverly Gray, author of *The Boreal Herbal*, an excellent book on wildcrafted medicine and food.

The wood of the plant is hard and has been used to make tool handles.

Saskatoon is important as a larval host for some of the swallowtail butterfly species. It is also food for deer, moose, elk, chipmunks, squirrels, bears, coyotes, and numerous birds. It got its name shadbush because it blooms when the shad fish (*Alosa* spp.) are at their tastiest and ready for catching (east coasts of Canada and the US).

Figure 25.5. Mule deer grazing on saskatoon.

Figure 25.6. Top and bottom of saskatoon leaf.

Caution

The leaves, bark, seeds, buds, and roots all contain cyanogenic glycosides that convert to cyanide when bruised or broken. Cooking or drying destroys the cyanide, rendering it harmless. Interestingly, the plants seem to have a higher content of the chemical in dry spells.

Similar species / look-alikes

- **Black hawthorn** (*Crataegus douglasii*) Small tree. Has large thorns on branches. Flower is smaller and petals are rounded. Berry is harder in texture and not as juicy. Seeds are large and light brown. They are oval with deep ridges. Edible. See black hawthorn chapter on page 100.
- **Oval-leaved blueberry** (*Vaccinium ovalifolium*) Shrub up to 2 m. Flowers are urn shaped and white. Leaves are small and oval. Edible.
- **Black chokeberry** (*Aronia melanocarpa*) Common cultivated garden plant in BC and parts of the US. Native to eastern parts of North America. Leaves are glossier and look more like apple leaves than alder leaves. They turn bright red in autumn. Flowers are white and small with rounded petals. Berry is dark purple or black when ripe. Ripen later in fall. The shrub is much smaller than saskatoon. Edible.

Figure 25.7. Black chokeberry.

Sheep Sorrel

Rumex acetosella | Polygonaceae

Synonyms

Fragile sour weed, red sorrel, field sorrel, common sheep sorrel, sour grass

Description

Perennial. Likes open spaces such as meadows and pastures. Forms colonies with its long skinny **roots**. Grows to about 30 cm tall when in flower. **Leaves** are arrow shaped with two distinct lobes at the base (hastate). Leaves form a basal rosette. They are juicy, sparkly, and dark green, turning a vibrant red in autumn. Plant is dioecious, meaning that plants are either male or female. **Male flower** has six sepals, no petals and six stamens. **Female flower** looks the same, but instead of stamens, it has a frilly white pistil. Plant is wind pollinated. Both types of flowers initially have green sepals that turn red as the plant flowers and sets seed. **Seed** is a papery achene. It is brown red and has three angles.

Figure 26.1. Sheep sorrel leaf. Note the two distinct lobes at base.

Uses

As a kid, I used to really like sour candy. This plant tastes just like that! The leaves are succulent and have a sourness to them that enlivens. I love the look on peoples' faces when they try this plant for the first time. First, they look at the plant skeptically as if they are not expecting to be impressed. One bite and their faces change into a smile, and they exclaim something like: "Wow that is a lot of flavors for a little leaf," or "Refreshing, where can I find more?" It has a lot of flavor, so when we use it, we do not need to use a lot. I love the leaves raw on sandwiches; they add a pickle-ness. It piques interest when added to a salad. I find it best raw. The leaves are at their best in early summer but can be eaten into autumn as well. They might be a little chewier as the season progresses. I sometimes even consume them when they are in their fall colors. Stunning! The plant loses some of its sourness when dried. It can be cooked with good results though. When lightly wilted in a frying pan and

Figure 26.2. Seeds and leaves.

Caution

Plants that contain oxalic acid should not be consumed in excess as this can interfere with calcium uptake in the body. Cooking lessens the oxalic acid. People with kidney problems should consume less of the plant or avoid it altogether. This plant can have a laxative effect. Avoid cooking it in pots made from aluminum, copper, or iron as the acid in it reacts with these metals.

added to fish dishes, it tastes great. The French make a traditional sorrel soup that tastes fresh and satisfying.

The reason for the sour taste is oxalic acid, same stuff that you find in rhubarb stalks. Too much of it can be detrimental to health. See caution box. Sheep sorrel contains vitamins A

Figure 26.3. Autumn leaf.

Figure 26.4. Sorrel flowers: female, *left*; male, *right*.

and C. Later in the season, I also harvest the seed spikes, which we add to bread or muesli or throw into a soup. They can be a bit dry on their own but are a source of protein and add some interesting texture.

Other uses

Sheep sorrel is a laxative in larger doses. Its puckering effect dries up mucus, so it is helpful for a sinus infection. Sheep sorrel is one of the ingredients in the Essaic tea that is used to

Figure 26.5. Wild seed loaf with sheep sorrel, sulphur cinquefoil, plantain, and white sweetclover seeds.

combat cancer. I am not sure why it is called sheep sorrel as it seems as if the plant can be toxic to sheep at times.

Similar species / look-alikes

- **Heartwing sorrel** (*Rumex hastatulus*) Grows in eastern and southern parts of North America. Grows taller (60 cm). Has a pronounced taproot. Its seeds are broadly winged; *R. acetosella* lacks the wing on its seeds. Can be used in the same manner.
- **Garden sorrel** (*Rumex acetosa*) Much larger plant (60 cm) with larger leaves that are arrow shaped rather than hastate. This plant is commonly cultivated in vegetable gardens. Edible.
- **Curly dock** (*Rumex crispus*) Leaves are longer and lack the flared lobes. The texture of the leaf is leathery. A much larger plant (1 m tall when flowering) that will have brown seeds with wings at maturity. Edible. See curly dock chapter on page 104.

Thimbleberry

Rubus parviflorus | Rosaceae

Synonym

Redcaps

Description

Deciduous shrub (up to 2.5 m tall). **Flowers** are large with five white petals and five green sepals. In the center of the flower, there is a rounded dome that looks like an underripe berry. This is covered with yellow pistils (female organs) surrounded by yellow stamens (male organs).

It is pollinated by bumblebees, bees, and other insects. Each pistil will develop into a **seed** inside a drupelet; many of these together form the aggregate **fruit** that looks like a raspberry. It is darker in color, and the drupelets are smaller and flatter than a raspberry. It looks like the name implies, a thimble. The **fruit** is deep matte red with small yellow hairs when ripe.

Canes/stems and branches **lack prickles** unlike their relative, raspberry. Like wild raspberry, it will bear fruit on second-year canes (floricanes). These have brown/grey bark that flakes. The first-year canes (primocanes) are initially green and bendy and will have no fruit or flowers the first year. **Leaves** are large, soft, and fuzzy to the touch. They are palmate with five lobes. The edges of the leaves are jagged. Their color is bright green early in spring and dull dark green in midsummer. Plant grows from **rhizomes** and can form large colonies in open clearings.

Rubus is the Latin name for "blackberry" and originally came from *ruber* meaning "red." *Parviflorus* means "small-flowered." This is a peculiar choice of name considering that thimbleberry has some of the largest flowers of all the *Rubus* species.

Uses

The thimbleberry patch is a popular destination for the little ones this time of year. The berries ripen successively over a period, so there is always some to snack on. Most of the berries

Figure 27.1. Thimbleberry flower.

Figure 27.2. Aggregate fruit of thimbleberry.

Figure 27.3. Large, soft, palmate leaves.

Figure 27.4. Berry looks like a thimble.

Figure 27.5. Structure after berry has been picked.

Caution

Use only fresh or completely dried leaves. Partly dried or wilted leaves can cause nausea and vomiting. The tannins in the plant can interfere with absorption, so be sure to leave time between ingesting thimbleberry leaves/shoots and any medicine or supplements you might take. The astringency of tea made from the leaves can cause nausea if too much is drunk on an empty stomach. Excessive use for a long time can irritate the bowels.

go straight into our mouths, only a few come home with us to be turned into jam. By the time they get home with us, they almost look like jam already, they are that mushy! Other berries are ripe at the same time, and we mix them together and get interesting jams. You can use the berries in any way that you would use regular raspberry. The berry is packed with good stuff like vitamins A and C, calcium, potassium, and iron.

Earlier in the year (April-May), we harvest the shoots of the shrub. When they first come up, they are soft, juicy, and green with a reddish hue. They can be snapped off in a similar manner to asparagus. Some of them are very astringent and pucker up your mouth. Others are mild and palatable. The shoots, boiled for five minutes and then stirred in butter, are a favorite side dish in our home. They can also be added to any meal that you cook; they shine in stir-fries. I like eating them raw in small quantities too.

The leaves can be eaten as well. Earlier in the year, they are better chopped up fresh and added to food; later in the year, they can still be used but are better if they are dried and then added to baked goods. They add a freshness to any meal. The flowers look pretty in a salad, but they don't have much flavor. Better to leave them to turn into berries!

Other uses

The leaves are large and soft and make excellent toilet paper. A tea made from the leaves is helpful in combatting diarrhea. I find that the tea can constrict the stomach, and it makes me constipated if I drink too much. The tea can also tighten up the skin and be used as a face wash. It dries out the skin somewhat so is best used on oily skin. It can be used as a wound healer.

Figure 27.6. Early spring shoots of first-year thimbleberry cane.

Figure 27.7. Sharing the bounty with the bears.

Remember that this plant is a favorite of the bears too. Make sure that there is plenty left for them and that you do not startle a bear in the berry patch. They blend in and hide perfectly under the large leaves. Once when I was moving too quietly through the berry patch, I came within meters of a young black bear. He/she was not sure what to make of me. We respectfully observed one another for several minutes and then went our separate ways. I was grinning from ear to ear. What a wonderful experience! My heart was beating just a bit faster.

Similar species / look-alikes

- **Purple flowering raspberry** (*Rubus odoratus*) Eastern species, popular in gardens. Large purple flowers. Perennial stems. No prickles. Palmate leaves with 3 to 7 lobes. Sometimes called thimbleberry. Can be used in the same way.
- **Salmonberry** (*Rubus spectabilis*) Found on the West Coast from Alaska to California. Taller shrub (4 m). Leaflets of three with jagged edges. Berries are lustrous and their colors vary from bush to bush. The berries can be yellow, orange red, or dark purple. Flowers are pink/purple. Perennial canes. Has prickles. Edible.
- **Wild red raspberry** (*Rubus idaeus*) Has fine prickles on the stalks and branches. Leaves are pinnately compound with 5 to 7 leaflets. Edible.

Figure 27.8. Wild red raspberry.

Oregon Grape

Mahonia aquifolium | Berberidaceae

Synonyms

Hollyleaved barberry, Oregon holly grape, Oregon holly, mountain grape, tall Oregon grape, mountain holly, mahonia

Description

Shrub (up to 3 m) with evergreen leaves. **Stem** is hard and gnarly with rough grey bark. **Inner bark** is bright yellow. **Root** looks like the stem. It is very hard and even more intensely yellow under the bark of the root than the stem. **Leaves** are orange/red when new, turning to a bright green. After this stage, they darken to a deep green, and during fall they become vibrantly red. Leaves are pinnate (5 to 9 pairs of leaflets) with spiky edges and a single vein. Leaf is hard and leathery at maturity and soft when young. The shiny yellow **flowers** are borne in racemes. To me the flower looks like a miniature daffodil. Flower has six large yellow sepals and six petals that look like they are fused together at times. It is pollinated by insects. **Berries** ripen in late summer-fall and are pale blue with a whitish coating; inside they are dark purple/blue and juicy. The berry has shiny red/brown **seeds**, shaped like boats.

The name *Mahonia* is in honor of nineteenth-century horticulturist Bernard McMahon; *aquifolium* means "sharp-leaved."

Uses

This is such a fun plant to harvest berries from. They hang in lovely clusters and transport well too. They are an excellent berry to harvest with kids. Since they are quite tart, the kids are unlikely to eat the entire basket before we can get home. We use them to make jelly, syrup, wine, and cordials.

Figure 28.1. Oregon grape flowers in May.

Figure 28.2. Leaves in midwinter.

Figure 28.3. Berries in late July.

Jelly made from Oregon grape and black hawthorn (*Crataegus douglasii*) is my absolute favorite. The berries can also be used as a sourdough starter. Throw a handful in a bowl with flour and lukewarm water. Let sit for a day in a warm place, and you will see it bubble away due to the wild yeast. The bread is mildly sour and does not rise a huge amount. My whole family enjoys the taste of it.

In late spring-early summer, the young leaves can be eaten too. Initially they are orange/red, and they look like last year's leaves. They also appear to have spikes on the edges. They try very hard to not look appetizing. Do not let this fool you! They are soft and juicy and sour tasting. Quite a treat! We pick them and add them to salads or sandwiches. They make a great trail nibble. They can be cooked too but do not hold their texture very well. They can be eaten until they stiffen and become too leathery and spiky to eat. The edible flowers have a sour taste like the leaves but with an added pollen flavor. We only eat the flowers sparingly because the berries are such a treasure that we want to make sure that as many flowers as possible turn into berries.

Other uses

The stems, branches, or root can be used for medicine. Since taking the root destroys the plant, I always opt for harvesting the stems or branches instead. The stems can be harvested all year, but some sources say that the potency is higher in the fall. The chemical responsible for the yellow inner bark is berberine, which helps to combat bacterial infections. See caution box. If you feel like your stomach is a little off and you suspect it was something you ate, a drop-

Caution

This caution mainly concerns the root and stem of the Oregon grape. Avoid using if pregnant; they may stimulate the uterus. Do not use while breast-feeding. Do not use for newborns or very small children. Exposure to berberine for the fetus and newborn infant with jaundice can cause kernicterus (brain damage). It is very rare and preventable. Avoid altogether for liver or kidney problems. Some pharmaceutical medication might react negatively with Oregon grape, so consult with doctor before using.

per full of Oregon grape tincture can help your digestive system combat whatever bacteria you ingested.

We use it as a bitter to stimulate appetite and help the liver do its job. Either a decoction or a tincture works, as long as you can taste the bitterness of it. I feel that the decoction helps with digestion, elimination, and to clear up the skin. The decoction or tincture of the plant can be added to creams, salves, and lotions and used externally on psoriasis, acne, and eczema. Try a small patch first to see how your skin reacts to it. You can also use the decoction externally to wash any wounds. The strong microbial action of the plant will stop a bacterial infection.

To make a decoction, simmer the stems for about 25 minutes. Sometimes you can get a couple of batches of decoction out of the same stems. Lower stems tend to be more yellow, and they are stronger medicine. Make sure that your

Figure 28.4. Seeds and berries.

Figure 28.5. Oregon grape sourdough bread.

Figure 28.6. Young spring leaves.

Figure 28.7. Inner bark is bright yellow.

Figure 28.9. Falsebox.

Figure 28.8. Oregon grape wreath.

stems have a dark yellow inner bark and that your decoction turns a rich orange/brown. I do not drink this beverage for more than two weeks in a row and usually only one cup a day. I might give the kids one tablespoon once a day for 3 or 4 days in a row and then have an extended break. Because the plant is so potent, I tend to harvest very little of it each year. My dried stems have been good for about two years in proper (cool and dark) storage.

Another thing we use the Oregon grape for is to make wreaths for holidays, as gifts, or as part of ceremonies such as remembering lost ones. The leaves are very durable, and the dark red leaves in the winter make a great contrast to dried grasses, twigs, and green foliage such as cedar. Oregon grape is used in commercial floral displays, and indiscriminate picking of the leaves has caused bushes to decline in some areas. If you are using this plant, please pick only a few branches from each bush and cut them carefully to avoid injury to the plant.

Harvest your medicine (stems) the same time as your decorating material (leaves), and nothing on the branch will be wasted.

Similar species / look-alikes

- **Dwarf Oregon grape** (*Mahonia nervosa*) A smaller Oregon grape that can be found on the West Coast. Lower, more ground-hugging plant (10 to 60 cm, can reach 2 m at times). Leaves are pinnate with 9 to 19 pairs of leaflets. The leaflets are less shiny than *M. aquifolium,* and their mid-vein is less pronounced. Can be used in the same way.
- **Creeping Oregon grape** (*Mahonia repens*) Ground hugging (4 to 30 cm). Leaves are blueish, dull, and more rounded than *M. aquifolium.* Leaves are pinnate with 3 to 5 pairs of leaflets. Mid-vein is less pronounced than *M. aquifolium,* and the spikes on the edges are less pointy. Can be used in same way.
- **Christmas holly** (*Ilex aquifolium*) Can grow tall (up to 25 m). Leaves are alternate, shiny green, and spiky. Flowers are white; berries are red rather than blue. **Berries are toxic.** Plant has some medicinal uses. Exercise caution.
- **Falsebox** (*Paxistima myrsinites*) Low-growing shrub (up to 60 cm). Leaves are evergreen and small (1 to 3 cm). Leaves are nearly opposite. Flowers are maroon, and the fruit is an oblong capsule. Used medicinally.
- **Virginia creeper** (*Parthenocissus quinquefolia*) Vine with palmate leaves with five leaflets. Leaves have jagged margins. Flowers are small and greenish. Berries hang in clusters from bright red stalks. Berry is blue black and round with a whitish coating. Stems have adhesive pads that assist in climbing. False Virginia creeper (*P. vitacea*) is identical, apart from the lack of this adhesive pad. Both are **poisonous to humans**.

Figure 28.10. Virginia creeper, *left*; Oregon grape, *right*.

Black Hawthorn

Crataegus douglasii | Rosaceae

Figure 29.1. Hawthorn berries in July.

Figure 29.2. Flower in May.

Synonyms

Douglas thornapple, black hawberry, western thornapple, Douglas hawthorn, river hawthorn

Description

Native small tree/shrub (up to 8 m) with grey flaky **bark**, often covered in lichen. Branches have large black **thorns** (1 to 3 cm). **Leaf** is oval with jagged points along the entire leaf and 5 to 9 small lobes on tip. Leaf is dark green on top and whitish below; it is quite leathery. **Flower** is white, has five round petals, four orange pistils and approximately ten pink stamens. Flowers are borne in flat-topped clusters at the tips of the branches. They have a musty odor that attracts midges and other insects that pollinate them. **Berries** (pomes) are dull black or dark purple when ripe. At their base, they have a five-pointed star remaining from the flower structure, just like an apple. Berries are mealy and contain 4 or 5 wedge-like **seeds** stuck together to form a large ridged oval.

Crataegus refers to the hardness of the wood. *Douglasii* refers to David Douglas who was a nineteenth-century Scottish plant collector.

Uses

We harvest the berries in the summer. They are sweet but can be hard to eat due to their large seeds. It is easier to process the berries into jelly, syrup, or wine. They are high in pectin so are great at thickening and setting jellies. Their

Figure 29.3. Top and bottom of leaf.

pectin content decreases the more they ripen. For jellies and fruit leather, it is best to use the just-ripe berries. They have developed their sweetness, and they have a good amount of pectin.

One way to preserve them is to make a thick paste by squeezing the berries with a little added water through a mesh bag and then letting the paste dry in a skinny, flat sheet that will turn into fruit leather. I like to dry my fruit leather in the food dryer, but anywhere with good ventilation and protection from flies works. It will harden and last for a long time in cool and dry storage. You can add other berries to the mixture too. The thinner you make it, the faster it will dry. This stuff is great to take along on a hike or send with kids in their lunchboxes. I also like to dry a few of the berries (seeds and all), and these I throw into a pot of tea with herbs to add a sweet fruitiness.

Earlier in the year, we collect the leaf buds and add those to our salads and sandwiches. We also collect the leaves and flowers to dry for tea.

Other uses

Hawthorn is known as a gentle heart medication. It allegedly both slows down an overexcited heart and increases the rate of a sluggish one. It can also help with high blood pressure and lowering of cholesterol. I like to use it as a tonic (a plant that will restore good health and that you can take for a longer period). The berries are safe if you do not consume the seed. The seed contains cyanogenic glycosides, just like apple seeds and cherry pits. The tea made from the leaves can be taken long-term if dried before consuming to remove small amounts of cyanide in them. I drink hawthorn tea if I feel heartache from grief or anxiety. You know the feeling...when your heart is racing and you feel kind of panicky, just wanting to get away from the intense emotions you are experiencing. This tea can help calm the heart and mind down. I mix the flowers, leaves, and berries and steep them covered for about 15 minutes. I add honey and drink it hot or cold.

Figure 29.4. Hawthorn berry seeds.

Figure 29.5. Spring leaves on stem with thorn.

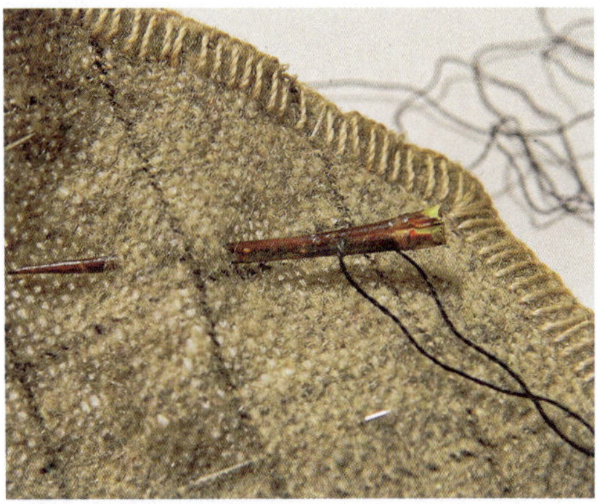

Figure 29.6. Sewing needle made from thorn.

Figure 29.7. Mourning cloak butterfly.

The formidable thorns of the hawthorn tree can be used as sewing needles. They are quite strong but make a large hole in the cloth. I make a hole or slit at the larger end of the thorn and then thread it and use it as a regular sewing needle. The kids and I have made fishing hooks with the thorns too, but we have yet to catch any fish with them. We have not been able to catch any fish with conventional fishing hooks either, just to give you an idea of our level of expertise.

"Haw" is an old word for "hedge," and hawthorn was used in Europe as hedges to divide land into plots. It works well for this purpose with its thorns and shrubby appearance. The hedges also create great habitat for small birds to build safe nests in. The black hawthorn is a native plant in North America and as such is a great plant for land restoration; its spreading roots can stop erosion.

On our property, the berries are popular with the black bear. The tree is a larval host for the beautiful mourning cloak butterfly (*Nymphalis antiopa*).

Similar species / look-alikes

- **Piper's hawthorn/red hawthorn** (*Crataegus chrysocarpa* var. *piperi*) Shrub. Has thorns. Berries are red. Ripens late in the season. Everything else is almost identical to *C. douglasii*. Used in same manner.

Caution

Beware of the large thorns. They can do serious damage. Do not eat the seeds as they contain cyanogenic glycosides that break down and release cyanide. Can cause nausea, vomiting, and diarrhea.

Figure 29.8. Underripe Piper's hawthorn, *left*; ripe black hawthorn, *right*.

- **Downy hawthorn** (*Crataegus mollis*) Shrub/small tree. Has thorns. Berries are red. Leaves are toothed around the entire margin and are slightly hairy. Used in same manner.
- **Common hawthorn** (*Crataegus monogyna*) Shrub/small tree. Has thorns. Introduced from Europe. Berries are red. Leaves are deeply lobed. Used in same manner.
- **Common buckthorn** (*Rhamnus cathartica*) Introduced shrub with thorns. The leaves are oval with finely serrated edges and arcuate leaf venation. The berry is black and round. It lacks the star shape. The berries grow on the stalk rather than on the tips of branches. The flowers have four petals and are green. **Poisonous.**
- **Blackthorn/sloe** (*Prunus spinosa*) Introduced shrub. Large thorns. Blue to black rounded berries that lack the star shape. Coated with whitish powder. Grow on stalk rather than in clusters at tip of branches. Leaves are small and oval with jagged edges. The white flowers bloom before the leaves emerge. Edible.
- **Saskatoon** (*Amelanchier alnifolia*) Shrub/small tree. Berries are blue. Leaves are rounded with a few jagged teeth at the tip. Leaves are blue green. No thorns. Edible. See saskatoon chapter on page 84.
- **Black chokeberry** (*Aronia melanocarpa*) Small bush. Commonly cultivated garden plant in BC and parts of the US. Native to eastern parts of North America. No thorns. Leaves are glossy and look like an apple leaf. They turn bright red in autumn. Flowers are white and small with rounded petals. The berry is dark purple or black and looks like a small apple. Edible. For picture see page 87.
- **Chokecherry** (*Prunus virginiana*) Small tree. No thorns. Small white flowers borne in racemes. These later turn into large hanging clusters with fruit (drupes). The shiny fruits are dark red to purple to almost black with one large stone pit. They do not have the star shape like an apple. Leaves are oval and have serrated margins. Edible. See chokecherry chapter on page 113.

Curly Dock

Rumex crispus | Polygonaceae

Synonyms

Curled dock, yellow dock, sour dock, bitter dock, bloodwort, coffeeweed, garden patience, narrow dock, out-sting, winter dock

Figure 30.1. Spring curly dock leaves.

Description

Perennial. Starts off as a basal rosette of leaves. **Leaves** are long and linear with a distinct vein (often reddish) in the middle. Texture of the leaf is soft when young and later stiff/leathery. The color is dark green with wavy edges. As the leaf gets older, it often develops brown-red stains. Petiole is almost as long as the leaf. Later in the year, a ribbed **stalk** (1 m) shoots up that holds smaller leaves, flowers, and then seeds. Around the base of each branch, there is a papery sheath called an ochrea. **Flowers** are tiny and reddish green. Flowers have three inner sepals, three outer sepals, six stamens, three styles, and one ovary. The flowers are wind-pollinated and lack scent. **Seed** is inside a pyramid-shaped papery husk. Seen from one side, it looks like a heart. On each side of the seed husk, there is a lighter-colored mound. The seed inside is glistening brown and has three distinct sides. Each husk carries one seed. Initially the seeds and husks are green and then turn a rusty red at maturity. **Root** is a large yellow taproot.

Rumex is the Latin name for "dock and sorrel plants." *Crispus* means "curly" and refers to the edges of the leaf.

Uses

In August the seeds of the dock start to turn brown. I let them mature completely on the stalk, and then I snap off the stalk. I let the seed-covered stalks dry further in a place with good ventilation and out of the sun. Before the seeds start to loosen, I plunk it headfirst into a paper bag and store it like that until I have time to garble it ("garble" means to sift through dried herbs to make sure that your end product is of the highest quality). For the winter, I store the seeds in jars. There is no need to remove the outer husk of the seed. This part can be dried with the seed and eaten in the same way.

Sometimes I use the seed and husk whole; at other times, I grind them into a fine powder, depending on how I plan to use them. The flour turns your bread a dark color. The seeds are great added to granola, power bars, muffins, porridge, and bread. I also like adding them to soups and casseroles. You can eat the immature green seeds too. I cook them when they are fresh, but I do not dry them for the winter. I find it better to wait for the more nutrient-packed ripe seeds. You can roast the seeds and grind them to use as a coffee substitute too.

The new leaves can be eaten raw in the spring. They are deliciously sour and taste great in a salad, on a sandwich, or made into a dip with sour cream. They are similar to sheep sorrel (*Rumex acetosella*) in flavor. The older leaves can also be eaten, but they can sometimes be too fibrous and bitter to be enjoyable.

In the springtime in Armenia, people harvest the dock leaves and braid them. They then hang the braids in a windy place out of the sun to dry. During the drying process the braid ferments slightly, which gives the herb a distinct flavor. These fermented leaves, called aveluk, are then used in soup.

The young shoots/stalks can also be eaten. They need a little bit of peeling and should be

Figure 30.2. Mature seeds on stalk.

Figure 30.3. Seed and seed husk

Figure 30.4. Seed bar with dock seeds, hazelnut catkins, and nettle seeds.

Figure 30.5. Dock roots, note yellow color.

used before they turn woody. The window for eating this part of the plant is quite small. They taste great and refreshing. I eat them both raw and cooked.

Other uses

The root of dock is bright yellow and very bitter. I usually harvest my root in the fall. I make sure that I do not harvest too many to not endanger the plant abundance. Curly dock is an introduced species and can be quite weedy, but here on the farm, I feel we have too little of it. I like chopping the root into small pieces while it is fresh and then dry it. It becomes very hard to chop once it has dried. I add the dried pieces to vodka, turning it into a dark yellow tincture, which is beneficial as an iron booster for people with anemia. I find it helpful to take a spoonful of the tincture if I have a heavy period and feel lethargic. This boosts my energy level and my stamina. I like mixing it with stinging nettle (*Urtica dioica*) for added benefit. The root can also be used as a bitter to improve digestion and appetite, and be made into a tincture or decoction. Taking a dropper of the tincture a few minutes before a meal gets the bile juices flowing and aids in proper digestion. Oregon grape root/stalk (*Mahonia aquifolium*) can be added. It provides further bitterness as well as protection from pathogens.

Caution

Contains oxalic acid. If eaten in too large quantities, it can hinder absorption of minerals in the body, especially uptake of calcium. Cooking removes some of the acid. Avoid this plant if you have kidney problems. Can exacerbate conditions such as arthritis. Avoid during pregnancy and breastfeeding.

Similar species / look-alikes

- **Narrow-leaved dock** (*Rumex triangulivalvis*) Native North American dock species. Leaves are much skinnier, and the plant is daintier than *R. crispus*. Edible.
- **Yard dock** (*Rumex longifolius*) Introduced from Eurasia. Almost identical to *R. crispus*. The best way to tell them apart is by looking at seeds and husks. The seed husk of *R. longifolius* is larger, and it lacks the rounded mound on each side of husk. Edible.
- **Round-leaved dock** (*Rumex obtusifolius*) Leaves are more rounded. Seed husk has spiky edges rather than being heart shaped. Edible.
- **Garden orach** (*Atriplex hortensis*) Leaves are heart shaped rather than linear. Seed husks are flattened rather than triangular. Seen from the side, the husks are round. Seeds are black and round. Edible.

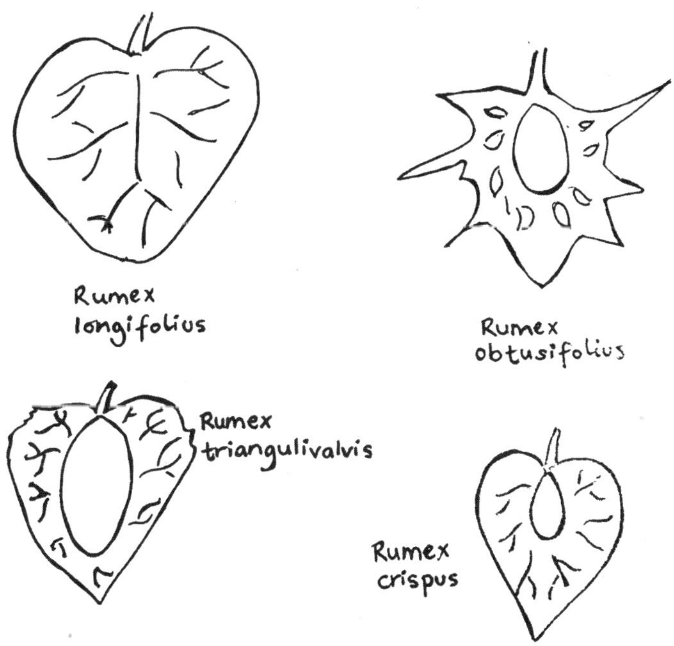

Figure 30.6. Seed husk comparison top left to bottom right: *Rumex longifolius*, *Rumex obtusifolius*, *Rumex triangulivalvis*, and *Rumex crispus*.

Western Giant Puffball

Calvatia booniana | Lycoperdaceae

Synonym

Boone's puffball

Description

Gasteroid fungi (bears its spores inside its fruiting body). Large round puffball (20 to 60 cm across) has **polygonal raised plates** on its surface that flatten as mushroom gets older. The **outer skin** (peridium) comes off easily to reveal the inner **gleba** (main meat part of mushroom that on maturity will become the spores). The peridium is initially white but turns brown or tan with age. The gleba on young specimens is bright white and spongy, reminiscent of marshmallows. As it matures, it will turn first gooey and yellow, then olive, and eventually dark brown and spore-filled. When the **spores** have all formed, the peridium of the mushroom will be stretched, and the polygonal raised plates disappear. The puffball will eventually crack to release the spores. Western giant puffball **does not have a subgleba** (a sterile part beneath the gleba that will not contain spores later. See drawing on page 112). The attachment to the ground is a small knob and it easily dislodges from the ground. **Spores** are olive-brown, and the mushroom smells a bit funky when ripe. Likes to grow in fields and sage brush areas.

The name *Calvatia* means "bald object" in Latin. *Booniana* is in honor of botanist William Judson Boone.

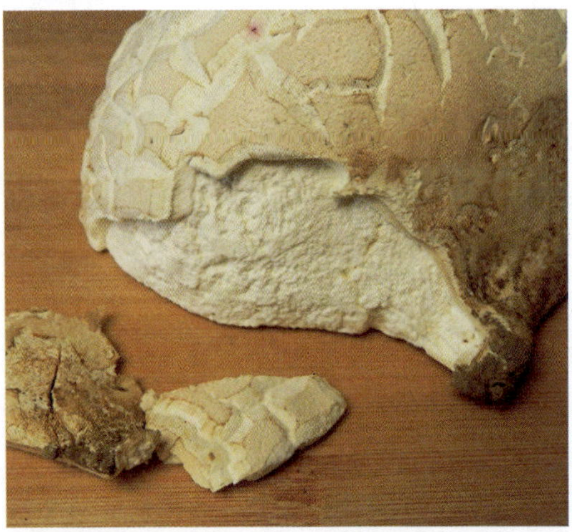

Figure 31.1. Note the knob where the puffball attaches to the ground.

Uses

This mushroom shows up in the same exact spot each time it makes an appearance at the farm. However, it cannot be relied on to come every year, so when it does show itself, we rejoice. The kids exclaimed: "Mom, one of the chickens has escaped!" Looking from afar it did indeed look like one of our large white chickens a way out in the field. Because I remembered that we in previous years had found mushrooms in that spot, I danced with joy and hurried the kids across the field for a rare sighting of a giant puffball. It is an experience you will never forget. This mushroom is just so unbelievably large. There was a recorded giant puffball in 1877 in New York State that measured 162.5 cm in diameter, and it was mistaken for a sheep! Ours was nowhere near that size, but when we got out there, we discovered a whole fairy ring of them. The kids thought the mushrooms looked just like soccer balls and wanted to kick them. I had to hold them back and explain that some of these mushrooms were going to be food and the rest left to release their spores.

They were in perfect condition for eating. The inner portion (gleba) was bright white and not attacked by any grubs or worms. As soon as even a portion of the gleba changes color from white to yellow, the mushroom can no longer be eaten. The texture becomes gooey, and they

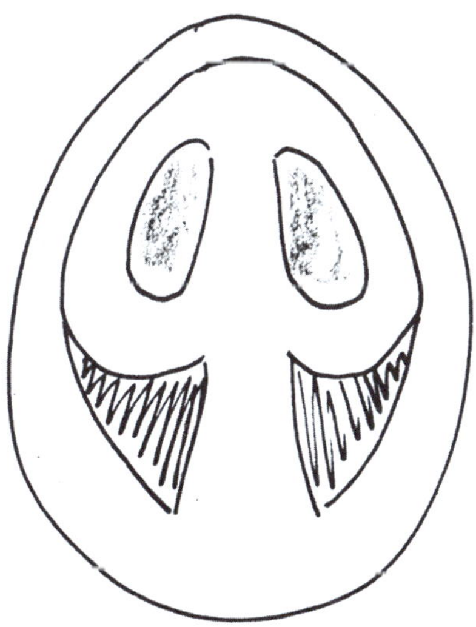

Figure 31.2. Potential poisonous look-alike: a young amanita. When the *amanita* "egg" is cut in half from top to bottom, a newly formed toadstool mushroom can be seen inside. The mushroom will later push out from the "egg," the stalk will elongate, and cap will flatten. The egg becomes the volva (cup-like structure at the bottom of the stalk).

Caution

Can give digestive upset to some individuals. Can cause gas if eaten in large quantities. The related species *Calvatia gigantea* can affect some adversely. This manifests in nausea, vomiting, and diarrhea a couple of hours after eating. Usually, full recovery is within hours. It is possible for some individuals to experience the same with *C. booniana*. With any mushroom, try only a little the first time you eat it to see how you react.

When puffballs are small, they can be mistaken for budding *Amanita* spp. **The *Amanita* genus includes some of the deadliest mushrooms.** Cut the mushroom open. If it is a solid white with soft flesh (marshmallow texture), you have a puffball; if there is an outline of gills and stalk, it could be an *amanita* and should be avoided. Wash knife before cutting any other food. If the flesh is slightly pinkish, it could be an immature earthball (*Scleroderma* spp.). Do not eat! This species is poisonous too. Both the *Amanita* (6.5 to 15 cm when mature) and the *Scleroderma* (2.5 to 10 cm when mature) are much smaller than the mature *Calvatia*, so they should not be easily mistaken for one another.

When foraging for mushrooms, always consult mushroom guides as they have the most comprehensive information.

Figure 31.3. Note the polygonal raised plates on the surface.

Figure 31.4. Bigger than her head!

Figure 31.5. Meal of sliced puffball.

taste awful. We brought our mushroom home and gently took off the outer skin (the peridium) and sliced the gleba in large slabs. These I fried in butter and added a bit of soy sauce and a splash of maple syrup. Served with rice and cooked vegetables, it tasted like tofu. It has a delicious texture and soaks up any flavor you cook it with, so it is a very versatile mushroom. After eating great quantities of it that night, we can now attest that it can give you gas if you eat too much in one sitting.

Figure 31.6. Sliced gleba (fruiting body).

Since you will have so much of the mushroom meat once you find a giant puffball, you will need to preserve some of it for the future. I like drying them in little cubes, which look like

Figure 31.7. Spore stage, not edible.

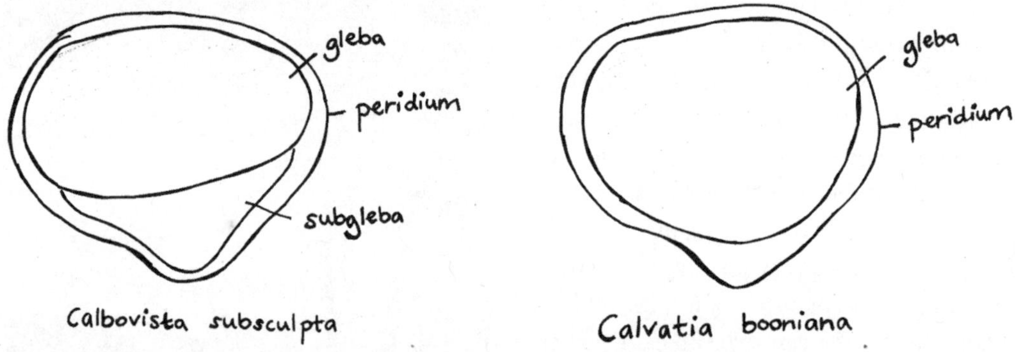

Figure 31.8. Comparison between *Calvatia booniana*, and *Calbovista subsculpta*. Note the lack of subgleba in *C. booniana*.

small marshmallows and feel like Styrofoam. I store these in my pantry, and when I want to use them, I rehydrate them for 10 minutes or so before chucking them into soups and stews. The cubes can also be ground into a powder that is easily added to any meal you want. You could also choose to preserve the mushroom by frying and freezing.

Other uses

In the excellent book *The Fungal Pharmacy*, Robert Rogers writes that the giant puffball (*Calvatia gigantea*), a close relative to the western giant puffball, has medicinal qualities. The spores were used to stop hemorrhaging effectively. Giant puffball (*C. gigantea*) contains calvacin, which can stop or slow down the growth of sarcoma; this mushroom is being researched as a tool to fight cancer. The large mushroom was used in the past to carry fire. A chunk of the mushroom was left to smolder and could easily be carried to the next spot and then reignited to start a new fire.

Similar species / look-alikes

- **Giant puffball** (*Calvatia gigantea*) Large white puffball (20 to 50 cm). Lacks the polygonal plates and is instead completely smooth. Used in the same manner. Edible.
- **Sculptured puffball** (*Calbovista subsculpta*) Large white puffball (8 to 15 cm). Looks very similar to *C. booniana*. Best way to tell apart is that sculptured puffball has a subgleba that western giant puffball is lacking. See drawing. Used the same way. Edible.

Chokecherry

Prunus virginiana | Rosaceae

Synonyms

Bitter-berry, black chokecherry, Virginia bird cherry, western chokecherry

Description

Native small tree (1 to 4 m) with grey-red **bark** with lighter-colored lenticels. It often has a black hard fungus (*Apiosporina morbos*) growing on the branches, also called black knot disease. **Leaves** are oval with small teeth around the edges and a pointed tip. They are green on top and white on bottom. Near the leaf on the petiole are two yellow-brown glands. **Flowers** are white with a yellow-orange center and grow in long clusters (racemes) near the tip of the branches. Each flower has five rounded petals, five sepals, one style, and numerous stamens. The protruding stamens give the flower clusters a fuzzy look. Flowers have a delicate scent like almond blossoms. They are pollinated by many kinds of bees and flies. In early fall, the **berries** (drupes) ripen, hanging in drooping clusters and varying in color from dark red/purple to black. They are round, shiny, and contain one large **seed** (pit). Stalks of the berry clusters turn pink. Berries are very astringent and pucker up your mouth, hence the name chokecherry. When the leaves are crushed or a twig broken off, the tree emits a bitter-almond scent that I find irresistible. It has a shallow, woody **root system** that spreads outwards by runners. This tree can at times form thickets.

In western Canada, you can find two varieties of chokecherry: Western chokecherry (*Prunus virginiana* var. *demissa*) with a

Figure 32.1. *Prunus virginiana* var. *demissa*.

Figure 32.2. Leaf bud in late March.

Figure 32.3. Flower raceme.

Figure 32.4. Western chokecherry next to black chokecherry.

burgundy-colored berry, and black chokecherry (*P. virginiana* var. *melanocarpa*) with an almost black berry. They can be used interchangeably. Many people call them both western chokecherry.

Uses

This year is a good year for chokecherry. The branches are hanging low with an abundance of fruit. We will feast on these berries as will the birds, rodents, and other large mammals in the woods. The berries can be eaten fresh. Care should be taken to spit out the seeds. They contain cyanogenic glycosides and can be dangerous if consumed in great quantities.

Both western chokecherry and black chokecherry work beautifully for making jelly, syrup, cordial, and wine. Without added pectin or fruits high in pectin, such as crab apple or underripe apples, the chokecherry jelly will not set. I make a thick syrup with chokecherry juice and underripe apples. It tastes wonderful on pancakes.

Other uses

I like to dry the berries to have a supply of them for winter. They work great added to cold and flu teas, giving flavor and beneficial nutrients. This syrup added to some hot water with a splash of rum is a nice beverage to ward off a cold. Herbalists use the bark to treat coughs and diarrhea. I have never used the bark myself so cannot comment on how well it works.

The wood is hard and can be crafted into items such as tool handles and canes. This pioneer species can be used to regenerate disturbed grounds and attract native fauna. Many animals

Caution

The seeds, leaves, and bark all contain cyanogenic glycosides in varying amounts. That is what gives them their bitter-almond scent. The seed contains the largest amount of it. Cyanogenic glycosides change into cyanide when broken down. Cooking or drying lessens the cyanide content.

Figure 32.5. Top of leaf. Note the two glands on stalk below leaf.

Chokecherry Syrup

12 cups berries
4 underripe apples
4 cups water
3 cups sugar

- Cook berries for 20 minutes until soft. Mash with a potato masher. Hang in mesh bag to drip overnight. I squeeze the bag to get as much of the fruit as I can, but omit that if you want a clear jelly. Discard seeds.
- Cook peeled and cored underripe apples separately.
- Combine chokecherry juice and apples in pot and add sugar. Stir on high for 10 minutes. Can or freeze.

Figure 32.6. Bottom of leaf.

and insects rely on chokecherry for food. Deer and rabbits graze on foliage and twigs. Numerous animals such as bears, raccoons, and birds eat the berries. Caterpillars and other insects eat the foliage.

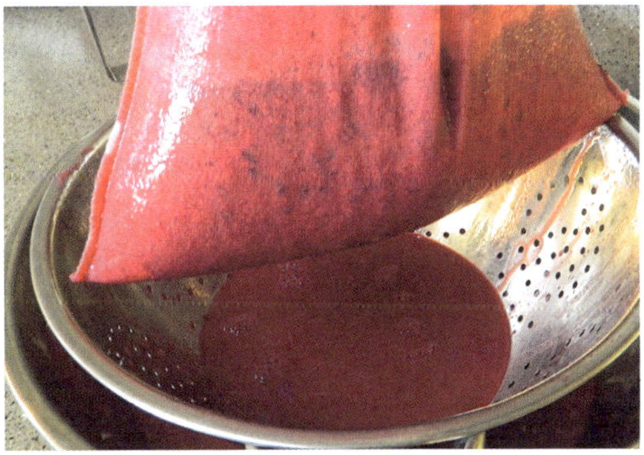

Figure 32.7. Chokecherry juice.

Similar species / look-alikes

- **Wild black cherry/rum cherry** (*Prunus serotina*) Larger tree (9 to 20 m). Leaves are skinnier, and the teeth on the edges are incurved. It is hard to tell the leaves apart unless you have them side by side. The flower raceme of *P. serotina* is slightly longer. The two species can be used interchangeably.
- **Pin cherry** (*Prunus pensylvanica*) Bright red berries hang in clusters with each berry stalk attached to the branch rather than to a raceme. Edible.
- **Saskatoon** (*Amelanchier alnifolia*) Blue berries look like little apples. The remainder of the flower structure forms a star at the bottom of the berry. Chokecherry berry (drupe) is in comparison completely smooth. Inside the saskatoon berry (pome) are multiple seeds rather than just one large stone. Edible. See saskatoon chapter on page 84.
- **Common privet** (*Ligustrum vulgare*) Introduced species from Europe. It is often used in landscaping to form hedges and has escaped into the wild. The best way to tell them apart are by the flowers. Privet has tubular flowers with four petals that are bent backwards. Their smell is unpleasant. Leaves are small with entire margins. They are arranged in opposite pairs on the branch. Berries are dark black and are borne in panicles rather than in racemes. Inside are 1 to 4 seeds. Berries are **toxic** to humans.
- **Black hawthorn** (*Crataegus douglasii*) Small tree. Has large thorns on branches. Flowers and fruits are in clusters at tip of branches rather than in racemes. Inside berries there are 4 or 5 wedge-like seeds. Edible. See black hawthorn chapter on page 100.
- **Chokeberry** (*Aronia* spp.) Small bush. Commonly cultivated garden plant in BC and some states. Leaves are glossy and look like apple leaves. Flowers are white and small with five rounded petals and grow in clusters at tip of branches rather than in racemes. Berry is dark purple or black and looks like a small apple. Each berry contains numerous small wedge-shaped seeds. Edible. See page 87.

Figure 32.8. Privet in bloom.

White Sweetclover

Melilotus alba | Fabaceae

Synonyms

White melilot, tree clover, honey clover, honey-lotus, white millet, bokhara clover, sweet clover

Description

Tall (up to 2.5 m) annual or biennial herb. Woody light green **stalk** with furrows or ridges. **Leaves** are green grey with three leaflets (trifoliate). Each leaf has a jagged margin all around. Leaves grow on the stalk in an alternate pattern with skinny stipules at the base of each leaf. The small tubular **flowers** (4 to 5 mm) are white and are borne on narrow racemes on the tips of the upper stems. Flowers have five united petals, five united sepals, ten stamens and one pistil. During midsummer the flowers are full of nectar and emit a sweet honey smell. White sweetclover is pollinated by various bees, bumblebees, butterflies, and flies. **Seedpod** is ovoid with a pointy tip and has a netted pattern. Inside are 1 or 2 smooth light-brown seeds. One plant can produce a staggering number of seeds (350,000), and the seeds can remain viable in the soil for eighty years.

The name *Melilotus* partly comes from the Latin word *mel* that means "honey." Lotus comes from the Latin genus name for "birds' foot trefoils." *Alba* means "white."

Figure 33.1. White sweetclover flower raceme.

Figure 33.2. Trifoliate leaf. Note the jagged pattern around the entire leaf.

Figure 33.3. Skinny stipules at the base of the petioles.

Figure 33.4. Fruit/seeds.

Uses

In August, I collect the unripened seedpods. They grow abundantly, and we can quickly gather plenty for a meal. It can be a bit overpowering to only eat white sweetclover, but it is nice in a mixture. I use the unripe seedpods fresh or dried. They have a lovely perfumy flavor and great texture. I use them in soups, casseroles, pasta sauces, or vegetarian burgers. Later, the dried fruit/seedpods can be collected with the seeds inside. These seeds can be sprouted and added to bread and baking. They can also be ground up and added to flour mixes for baking.

The leaves can be harvested earlier in the year and used both fresh and dried. I like to pick them before the plant flowers to get the juiciest leaves. When dried, they smell like vanilla

Figure 33.5. Green seed and vegetable patties.

Green Seed and Vegetable Patties

1 cup green unripe plantain seeds (*Plantago major*)
¼ cup unripe white sweetclover seedpods

- Mix together. I also add in 6 to 10 fresh agrimony leaves (*Agrimonia eupatoria*).
- Grate one small zucchini, 2 beets, and 1 large parsnip.
- Add 1 egg.
- Add a little bit of sorghum flour, or any kind of flour, to bind it together.
- Add chopped up garlic and green onion to taste. I used garam masala, salt, and pepper for spices.
- Mix all together and form patties.
- Fry in oil or butter.
- I served my patties on a slice of homemade bread with a cold creamy cucumber salad on the side. You can add any spices you would like or exchange the grated vegetables for others.

due to the coumarin in the plant. See caution box. I like to use the dried leaves for a tea and powdered to add to baking or mix with cream cheese. I prefer the dried leaves to the fresh leaves, which are slightly bitter.

The flowers are edible and are pretty when added fresh to salads. They can also be made into a syrup that tastes marvelous. Very sweet! The flowers can be nibbled on while out in the fields. They are ladened with nectar. Another interesting way to use them is to pan-fry them without oil until they are dry and crispy.

Caution

If any moisture remains in your dried plants, it can cause them to mold and turn the coumarin in the plant into dicoumarol; this chemical stops your blood from clotting. Moldy hay caused numerous cows to die, and further investigation revealed that the culprit was white sweetclover that had been allowed to become moldy. Treatment of this type of poisoning is vitamin K and possibly a blood transfusion. When dried properly, it causes no harm. Do not use this plant if you are taking blood thinners. Do not consume before surgery or childbirth.

Other uses

Herbalists have used this plant as a mild sedative and as an aid during bouts of anxiety. It has also been used to treat gas and indigestion. I can attest to its use as a relaxant, but I have not tried it for the other ailments.

Like other legumes, it fixes nitrogen in the soil and has been used as a cover crop to improve fertility of soil. It can be used as fodder for livestock, but there is always a risk of dicoumarol poisoning occurring. See caution box. The dicoumarol derived from sweet clover was originally used as a rat poison, but it did not work too efficiently. Later this chemical was used as a therapeutic anticoagulant called warfarin, approved for human use since 1954.

Figure 33.6. Yellow sweetclover in July.

Figure 33.7. Comparison: white sweetclover, *left*; alfalfa, *right*.

Similar species / look-alikes

- **Yellow sweetclover** (*Melilotus officinalis*) Very similar looking plant. Flowers are yellow and slightly larger (5 to 7 mm). The ridge pattern on the fruit is stripy; *M. alba* has a netted pattern on the fruit. Can be used in the same manner.
- **Alfalfa** (*Medicago sativa*) Very similar before it flowers. Alfalfa has a jagged margin only at the tips of the leaflets, whereas sweet clover has a jagged margin around the entire leaf. The flowers are blue/purple and grow in small clusters. The seeds are contained within spiraled seedpods. Edible.

Plantain

Plantago major | Plantaginaceae

Synonyms

Broadleaf plantain, white man's foot, greater plantain, common plantain, cart track plant, wild plantain

Description

Perennial herb. **Leaves** form a rosette that often hugs the ground. Leaves are oval with 5 to 9 distinct veins running lengthwise. If the leaf is pulled from the plant, the elastic **veins** often remain sticking out. They are harder to cut than the surrounding leaf. Leaf is a deep green; surface is shiny when young and matte when older. The fleshy petiole is almost as long as the leaf. **Flowers** are borne on a spike that towers above the leaves. Flowers are small and inconspicuous, brown green with yellow or purple stamens and one whitish pistil. They are wind pollinated. **Seedpods** are egg shaped or round. Initially they are green and later turn brown. **Seeds** are held in small brown husks that break into halves to spread the seeds. Plantain seeds are small,

Figure 34.1. Underside of plantain leaf. Note elastic veins.

Figure 34.2. Seeds and opened husk.

Figure 34.3. Seed spike harvest.

hard, black, and round. They are sticky when wet. Plantain has a fibrous **root**. This plant can handle compacted soil especially well and thrives in disturbed areas and on roadsides.

The name *Plantago* comes from the Greek *platus* that means "wide." *Major* means "large."

Uses

At this time of year, we harvest the seeds and seedpods. Both the mature brown spikes and the immature green ones can be found now. We use both, just in different ways. The unripe green spikes can be eaten like very small corn on the cobs. The kids call them "fairy corn." Throw the whole spike in boiling salted water for a few minutes and then eat it as you would corn on the cob. It takes some nimble teeth action to gnaw the seeds off the spike, and you need many to fill you up. It is a fun side dish. The unripe green seedpods can also be shaved off the spike and added to food dishes. They work well in casseroles, meat loaves, burgers, or soup. Check out recipe for seed and vegetable patties in the white sweetclover chapter, page 119.

I harvest the mature seedpods and seeds and let them dry upside down in a paper bag. I rub the dried spikes to loosen seeds and husks into the bag and discard the woody stems. I use the husks and seeds together in muesli, bread, muffins, or any other baking. The husks add nice bulk. It is the same stuff sold in health-food stores, under the name "psyllium"; a helpful dietary fiber. The seeds are packed with protein and add a nice crunch to your bread, reminiscent of a poppy seed. I gather a lot of these to use during the the year. This is one of the staples in our household.

In spring and early summer, we pick the new leaves, adding them raw to our salads or smoothies. Later, when they become more fibrous, it is better to cook them. They become mushy and make an excellent cooked green to add between layers of lasagna noodles.

They also make tasty chips when lightly oiled

and pan-fried for a few minutes. Another use for the larger leaves is to make wild dolmades. Make sure that you use young leaves; they can be too stringy when old. Fill the blanched leaf with precooked vegetables or meat mixture and roll up. These can then be cooked in the oven for about 15 minutes on 325°F (165°C) or simmered in a little bit of water and oil on the stovetop. Serve with a sweet-and-sour dipping sauce and sprinkle with roasted sesame seeds. Delicious!

Other uses

In our family, plantain is by far the most used plant for bee and wasp stings. I love how it empowers the kids to help themselves. When they get stung, they search the ground for the flat plantain leaf. They pull it off the plant, chew it into a wad, and apply that to the spot that hurts. All those actions together, with the expectation that this will work, coupled with the healing properties of the plantain leaf, enables them to help themselves in a situation that could otherwise be traumatic for a little one. Thank you, plantain! The leaf contains large amounts of vitamin K, and this helps to staunch blood flow. The plant is also an astringent that aids in healing the skin.

I like to make a smoothie-like liquid from plantain leaf mixed with avocado oil. I store this green goo in my freezer, and if I need a little extra face care (if those wrinkles are getting deeper or the moisture in my skin has disappeared from a day in the sun), I pull a chunk out and start to thaw it between my fingers as I rub it over my face. It dries on the face and feels weird and stretchy for a bit. After 10 minutes or so, I wash it off with lukewarm water, and the skin feels amazing afterwards. I think that I can

Figure 34.4. Plantain with flowers in July.

Figure 34.5. *Plantago major* flower close-up.

Figure 34.6. Breakfast buns with plantain seeds.

Caution

I have not come across many concerns regarding plantain. The Plants for a Future website mention that high doses can cause a drop in blood pressure, but I could not see any more information about that. Be mindful to drink more water when you consume the husks as they can be constipating on their own. People with obstructed intestines should avoid them entirely.

see a definite improvement in my skin from this simple method. I have also used it with effective results on patches of eczema.

The husks can provide dietary fiber and can help with both constipation and mild diarrhea. It does this by taking up water in the intestines, and the added bulk can help to move things along while the absorption of liquid can alleviate the diarrhea. One thing to keep in mind if using these husks is that you need to drink extra water to make sure that your intestines have enough liquid, otherwise constipation could worsen.

Similar species / look-alikes

- **Rugel's plantain** (*Plantago rugelii*) Native plantain species that is almost identical to *P. major*. To tell apart, look at the bases of the stalk: on *P. rugelii*, they are purple; on *P. major*, they are green. The seedpod is elongated/oval rather than egg shaped. Edible.
- **Narrowleaf plantain** (*Plantago lanceolata*) Leaf is narrow and lanceolate rather than

oval. Has prominent veins. Flower stalk is much shorter than *P. major*. Edible.

- **Hoary plantain** (*Plantago media*) Leaves are skinnier than *P. major*, and they have soft short hairs covering the entire leaf. Edible.

Making salves and body products

Conversation about plantain and wild plants with Terryn Corbett.

Terryn Corbett, owner and operator of Om Naturale, uses wild plants for her body products. She has a variety of soaps, creams, and body products available to her customers. With over a decade of making and selling these products in the Okanagan, she has become intimate with some of the local wild plants. Terryn has a background as an herbalist, and she half-jokingly says that she sneaks in herbs to heal your body. The plants in her products are there not just for their scent but also for their nourishing medicinal qualities; they will help your body heal itself. What becomes evident whenever you speak with Terryn is her deep respect and care for the plants in our environment. Her first concern is the well-being of the plants. As we speak about wildcrafting, she stresses the importance of taking only a little of a stand of plants, no more than 10 percent, so that enough is left for everyone. Because wildcrafting is becoming popular, she worries that even this method might not be enough to preserve stands.

Therefore, she has gone a step beyond and cultivates the wild plants she needs for her infusions. Plantain is a reliable perennial; it comes back in the same spot every year in her garden. We spoke about the various plants she uses in her products, and she echoes what many herbalists say. Become familiar with a few plants and get to know them really well, and you will find yourself with some powerful healing allies. You don't need to know every plant there is to use plants, you just have to learn a few plants very well.

Figure 31.7. Terryn Corbett, owner of Om Naturale.

When we speak about plantain, she enthuses how well it works for an itchy bug bite. Plantain leaf has natural antibacterial and anti-inflammatory properties that help heal wounds and take away itchiness on irritated skin. Over the years, Terryn has created many different salves for her customers; one that has been with

Figure 34.8. Some of the products from Om Naturale.

her from the very beginning is her "once bitten" salve. This simple yet effective salve has plantain leaf as one of its main ingredients. Terryn exclaims, "Plantain can't be improved upon, it is perfect on its own. I add it to a salve, and the plantain takes care of the healing." To Terryn and her family, grabbing a plantain leaf, chewing it up, and applying it to an itchy bug bite is second nature. By making a salve with plantain-infused oil, she has an easily applicable remedy year-round. She wants her products to make the art of using plants and their benefits available to people who might not be familiar with the world of plants. She knows that they work; she has experienced it herself. She doesn't want you to just take her word for it, she wants you to try it for yourself and see if it works for you.

Om Naturale are ground-breaking in that they pay close attention not only to what goes into their products but also to what their products go into. All products come in beautiful glass jars, reusable and environmentally friendly. As Terryn says, "You wouldn't put a beautifully crafted wine in a plastic bottle, so why would I want to put my natural, crafted with care products in plastic?" When I speak with Terryn, I am filled with hope for our future. She is consciously trying to make a difference with her business, emphasizing that we must nurture our community, both its plant and people. We can keep our community healthy by supporting local businesses that use sustainable practices. Terryn's motto is that she wants to leave the world a better place than when she entered it.

To learn more about Om Naturale and see what products they have available, visit omnaturale.ca.

Purslane

Portulaca oleracea | Portulacaceae

Synonyms

Green purslane, little hogweed, verdolaga, pursley, red root

Description

Annual herb that grows prostrate to the ground. Has fleshy hairless **leaves** (1 to 3 cm) that are paddle shaped. They are arranged in clusters/rosettes off the stem with smaller leaves overlapping larger ones. **Stalk** is thick and fleshy, sometimes with a reddish tint. **Flowers** are small and yellow with five petals that are notched at the top. They form at the tips of branching stems. When they bloom, they last for a few hours only. They also have cleistogamous (petal-less, self-pollinating) flowers. **Seedpod** is smooth, oval, and splits in half crosswise to form a little cup that holds the seeds for dispersal. **Seeds** are black, round, and flattened. **Root** is a deep taproot.

The name *Portulaca* means either "little door" from the Latin *portula* or "carries milk," referring to the plant's ability to stimulate lactation. *Oleracea* means "vegetable or herb" in Latin.

Uses

We harvest purslane from midsummer until late fall. After we harvested some of our root crops, the purslane took the opportunity to establish itself and spread out over the garden beds. It is doing us a great favor by holding the soil and retaining moisture. All aerial parts of purslane can be eaten. The succulent leaves and fleshy stalks are juicy and a little tart, with a hint of salt. The sour flavor is stronger in the morning.

We add it raw to salads and sandwiches. Sometimes the lower stalks can be a bit tough; if so, snip off your purslane greens a bit higher toward the tip, and you should have no problem. I love adding fresh purslane to a potato salad. It gives a tangy little surprise that makes a potato salad more interesting and less heavy.

Figure 35.1. Purslane flower in August.

Figure 35.2. Close-up of seedpod in September.

It works well in cold foods such as a quinoa or couscous salad. It mixes especially well in Greek yogurt and makes an interesting addition in a tzatziki dip.

Purslane can also be used in cooking; it is mucilaginous so will thicken soups and stews. Another fun way to use the stalks and leaves is to turn them into a refrigerator pickle. They do not hold up to regular canning processes so are best added fresh to a jar and covered with pickle brine. Set in the fridge for three to seven days before opening. It is recommended that refrigerator pickles be consumed within a month. In our house, there is no risk of them going bad. We gobble them up long before then.

The seeds can be gathered and eaten too. Here we do not get enough purslane to make that worthwhile, but I have read that in other parts of the world the seeds were ground up and made into seed cakes.

Caution

Contains some oxalic acid, which will lessen when cooked. If you have a history of kidney stones, be cautious using this plant since it may exacerbate the condition.

Other uses

Purslane has been used for a very long time by peoples all around the globe, for headaches, fevers, intestinal worms, dysentery, eczema, and diabetes to name just a few. Some uses in traditional Chinese medicine date back thousands of years. I like applying the juice of the entire plant to sunburn. It feels cooling, and it soothes the damaged skin. Purslane is a good source of vitamin A, iron, and potassium, as well as a high amount of omega-3, which has multiple benefits for the heart and the brain. It is not so common in green plants, more often found in fatty fish and nuts.

Purslane makes a great companion plant because it retains the moisture in the soil and its long strong roots can penetrate the ground and allow other plants' roots to tap into water and minerals at a deeper depth.

This plant can handle severe drought and is unusual in that it can utilize both C4 and CAM (Crassulacean Acid Metabolism) photosynthetic pathways. In the CAM process, the plant keeps its stomata closed during the day to reduce water loss. It appears as if purslane switches to CAM after a period of extensive drought. Plants like these might be of crucial importance as our climate changes.

Similar species / look-alikes

- **Spotted spurge** (*Euphorbia maculata/ Chamaesyce maculata*) Prostrate annual more common in eastern North America. Leaves are oval and grow in an opposite pattern. Often the leaves will have a darker spot on their top. Flower is white, and seedpods have three ridges. Stalk is hairy. The entire plant exudes a sticky milky latex that can cause skin irritation. **Poisonous.**
- **Ridge seed spurge** (*Euphorbia glyptosperma*) Looks very similar to spotted spurge but lacks the spot on the leaf. This plant is also less hairy than spotted spurge. Exudes milky latex. **Poisonous.**
- **Common knotweed** (*Polygonum arenastrum/Polygonum aviculare*) Prostrate annual, sometimes perennial. Leaves are alternate and oval. They are not fleshy like purslane. Has a papery silver ochrea where the leaf meets the stem. Flowers are green with pink-white edges. Edible.

Figure 35.3. Ridge seed spurge.

Figure 35.4. Common knotweed.

Common Burdock

Arctium minus | Asteraceae

Synonyms

Lesser burdock, little burdock, louse-bur, wild rhubarb, button bur, cuckoo-button

Description

Biennial plant; the first year it only creates leaves, and the second year it produces flowers and seeds (up to 2 m). Aggressive introduced species from Eurasia. Likes to grow in disturbed soil and can survive both sun and semi-shade. Its large **leaves** are formed like oval hearts, green on top and whitish on bottom. Bottom of the leaf and stalk have a fine fuzz. **Stalk** and petiole are often purple. Petiole of the lower leaf is hollow.

Figure 36.1. First-year common burdock leaves.

Plant forms pink-purple **flower** clusters at the top in its second year. Flower clusters are composed of pink-purple disk florets and floral bracts that are bent backwards, giving it that Velcro burr. Poking out of the disk florets are delicate white styles that divide in two, like the tip of a snake's tongue (bifurcated). These styles have a sheath of dark purple anthers around their base. **Achene (seed)** forms inside the **burr** that spreads by sticking to fur or people and hitching a ride to new places. Seed is flat, brown with spots and vertical ridges, and has small bristles at one end. **Root** is a large taproot.

Arctium comes from the Greek *arction* that means "bear." *Minus* means "smaller" and highlights the fact that it is smaller than greater burdock (*Arctium lappa*).

Uses

When you go out to grab your burdock root, you need to be equipped to battle. With enough begging, cajoling, and sweat, victory will be yours in the end. The taproot goes down deep and does not give up easily. Most often a portion breaks off and remains in the soil. The first-year roots can be used throughout the season but tend to be bigger and have more nutrients stored in them in the fall after the leaves die back. I harvest the root from the second-year

Figure 36.2. Underside of leaf. Note the velvety hairs.

Figure 36.3. Flower.

plant in the spring (before the flower stalk appears) to get the maximum amount of nutrients in the root.

We use the small or medium roots in stews, soups, and bakes. The larger ones I dry, grind, and use as coffee substitute or flour. On larger roots, I peel off the outer skin. Sometimes I soak the root for a few hours in baking soda. This step is not necessary, but it can take away some bitterness before you add the root to your cooking. I then thinly slice or dice the inner part and cook together with other domesticated root crops. Today's supper was carrots and burdock root sliced finely, fried with coconut oil and honey. It was delicious. The root contains potassium, calcium, iron, and amino acids. It is a good source of dietary fiber. It also contains inulin, which can cause gas.

We also make use of the stems of the second-year plant before the flowers appear. The window for using them is small (a couple of weeks in June where we are are). Once the flowers appear, the stalk turns woody and bitter. The stalk needs a tiny bit of peeling. It is crunchy and mild-flavored, and tastes great added to any meal. It can be used in a similar manner as broccoli stalk or celery. I have tried multiple ways to make burdock leaves palatable, but I find them insufferable. I have tried to use them in cooking, blah. I have boiled them in several changes of water and baking soda, still blah. Raw, blah. In an emergency I would eat them, but there are so many other nice parts of burdock to eat.

Figure 36.4. Stems.

Later in the year, the burrs will appear; most people are familiar with them as unwanted

Figure 36.5. Common burdock roots.

Figure 36.6. Burr in fall.

tag-alongs on clothes or dogs' fur. This plant's seed inspired Velcro, and the burrs really hold on. The seeds should not be eaten raw; they leave a tingling/numbing sensation that is alarming. This reaction might be because the seeds have tiny toxic hairs that easily get inhaled when consumed raw. I prefer the seeds roasted and ground into flour and used in bread. The seeds can even be harvested throughout the winter as they sway on dried stalks over the snow. Make sure you leave some for the birds who rely on seeds like this to survive the winter. You can sprout the seeds if you want, but the sprouts taste just like the leaves, bitter, bitter, bitter.

Other uses

Burdock has an abundance of health benefits. The root improves the function of the liver, making it easier for the body to get rid of accumulated toxins. I make a spring cleanse decoction with burdock root, milk thistle seeds (*Silybum marianum*), dandelion root (*Taraxacum officinale*), and nettle leaves (*Urtica dioica*) to remove any stagnant buildup in my body and get me ready for spring. The burdock root can help to clear up skin conditions such as psoriasis, eczema, and acne. It also helps with digestion and keeping you regular. The seeds have a blood-sugar lowering effect and can be useful in diabetes.

Caution

Do not use burdock in great quantities when pregnant; it can stimulate uterine activity. The inulin in the root can cause gas if eaten in large amounts. Do not eat the seeds raw as the bristles attached to them can be harmful to the airways.

Burdock/Sunflower Seed Coffee

- Dry burdock roots in smaller chunks. You can roast these, but it is not necessary.
- Dry sunflower seeds in husk. Fry these in a little bit of olive oil and a pinch of salt.
- Grind the dried burdock root and sunflower seeds together in a coffee grinder.
- Add 4 or 5 coffee beans if you want; this is not necessary, but I think it rounds it off nicely. Brew like you would ordinary coffee. I like it with milk or cream. This is a fabulous drink!

Figure 36.7. Seed. Note bristly hairs.

Figure 36.8. Comparison: young burdock (*left*) and arnica (*right*).

Similar species / look-alikes

- **Woolly burdock** (*Arctium tomentosum*) Can be recognized by its cobwebby flowerheads. I have not seen this burdock here, and I have never tried eating it. Edible.
- **Greater burdock** (*Arctium lappa*) Grows taller (3 m). Flowers are borne in open flat-headed cymes. The petioles of lower leaves are not hollow. Edible.
- **Cocklebur** (*Xanthium strumarium*) Leaves are slightly maple shaped. Flowers are white/pink. Male flowers form on top of plant on small balls, and female flowers form on an elongated burr further down the stalk. Has some medicinal qualities, but great care needs to be exercised since **some parts can be toxic**.
- **Heart-leaved arnica** (*Arnica cordifolia*) Heart-shaped leaf with a very distinct smell, reminiscent of calendula. Leaves have jagged edges. Can only be mistaken for one another in spring when both are small. Much smaller than burdock at maturity. Root is a rhizome. Yellow daisy-like flowers and no burrs. External use only. Can be **poisonous** if ingested.

Chickweed

Stellaria media | Caryophyllaceae

Synonyms

Common chickweed, winterweed, chickenwort, craches, maruns, starweed, satin flower, mouse-ear, stitchwort

Description

Annual that likes shaded moist areas. Grows in a low sprawling manner with decumbent tips. **Stalk** is bright green and juicy. If cut, it will exude **clear sap**—never white sap. A telltale sign for this species is that it has a **strip of hairs** on only one of the four sides of the stalk, alternating at each node. **Leaves** are small and oval with a pointy tip. Edges are entire, and the petiole has one strip of fine hairs. Leaves grow opposite on the branches. **Flowers** are white, borne at the tip of the stems, and have five deeply cleft petals. Petals are spaced apart from one another. Flower has up to ten stamens and a style that is parted in three. Sepals are green, oval, and covered in fine hairs. Sepals are as long as, or longer, than the petals.

Figure 37.1. Chickweed flower with five deeply cleft petals, making it look like it has ten petals.

Chickweed can form **roots** from lower nodes that touch the ground and can quickly form dense mats. **Roots** are fibrous and skinny. **Seeds** are small, round, and brown. They are held in cylindrical pods until released onto the ground. One plant alone can produce more than 25,000 seeds. It is pollinated by various small bees and flies, but it can also self-pollinate if it needs to.

The name *Stellaria* comes from *stellar* and refers to the star-shaped flower, and *media* means "midsize."

Uses

We use the entire plant of chickweed if it is not too fibrous. Usually, the tips of the plant are still good to eat no matter how old the rest of the plant is. From early spring to late fall, we can find chickweed on the farm in a tucked-away moist and shaded area of the garden. Nothing else will grow there with success, but the chickweed thrives. I thank my lucky star for that. I tend to this lush patch of chickweed with the same attention I do to my most prized tomatoes.

It gives us abundant food for a large portion of the growing season. My favorite way of eating it is fresh in a salad. It is juicy and gives great bulk. Just make sure you wash it well before tossing it into your salad or else you might find yourself chewing sand! It has a sweet and fresh flavor.

It can also be thrown into an omelet or quiche with good results. Once I made soup from it, but I added too many stringy old plants; my husband calls that experiment "the weed-whacker soup." If you intend to use it in the fall, make sure you only pick the top portion of the plant to avoid repeating my mistake.

Chickweed tastes great added as topping to pizza, mixed into a pasta sauce, or as greens on a sandwich. The plant can also be dried and used throughout the winter. I have not used it much in this way. I feel that the best part of chickweed is its juiciness, and it loses that through the drying process. It does not store for long as a dried herb and needs to be used up within a few months. A better way to store it is to make pesto with it and then freeze the pesto in portion sizes for a quick meal.

Figure 37.2. Note strip of hair on chickweed stalk and the hairy flower buds.

Other uses

Chickweed contains iron, copper, and vitamin C. It is beneficial for people with anemia. It also has some calcium, vitamins D and B, as well as a range of other micronutrients. Many herbals mention its use for the stomach and the bowels. The plant heals and calms down an inflamed stomach. Many sources also allude to its use to control weight. I find it helps my digestion and my overall feeling of well-being when I eat it.

I have used a water infusion of the plant on itchy skin or rashes with good results. It has a cooling effect and gives relief. I have also used an chickweed-infused oil on itchy skin, but that did not work as well as the water mixture. The oil was not cooling enough. The oil made a nice addition to a salad dressing though.

Chickweed is a great herb to use after you have been out in the sun all day and feel flustered and have a hard time settling. It will calm and cool. It can be taken internally as a tea or added to your bath to heal and calm your skin. I float the fresh plant in a mesh bag while I bathe. Although some sources speak of using it as a remedy for eczema, I have not satisfactorily treated eczema with it. I might not have used it for as long as needed or as consistently. With its abundant nutrients and minerals, it is a good herb to give to someone who is recovering from illness. Why not add it to your chicken noodle soup (just before serving) for flu-filled days?

Chicken-out Salad

Two handfuls of chickweed tips
One handful purslane leaves and stalks
3 large leaves of kale
½ cup chickpeas
1 nectarine
¼ roasted chicken breast

- Cut ingredients into bite-sized pieces and mix. Nutritious and delicious!
- For dressing, we used a chickweed-infused oil mixed with balsamic vinegar and spices.

Caution

Chickweed contains varying amounts of saponins. They are removed through cooking. I do not worry about these saponins. I eat the plant raw in great quantities, and I have never had any problems. Potentially it could cause diarrhea and nausea if consumed in massive amounts. Try a little bit the first time you eat it and see how you react before filling your whole plate.

Figure 37.3. Chicken-out salad.

Chickweed is loved by chickens (go figure!) and can be given as fodder to your fowls. It will also attract small wild birds; mourning doves, grouse, and sparrows enjoy the seeds of this plant. Rabbits, groundhogs, and deer like to munch on the foliage. Interestingly, the seeds of the plant can pass intact through the deer's stomach, thus dispersing it far from the mother plant. It is a lovely edible ground cover in a garden, and it retains the moisture in the soil. As it decays, it adds valuable nutrients such as potassium and phosphorus into the soil.

Figure 37.4. Mouse-eared chickweed.

Similar species / look-alikes

- **Scarlet pimpernel** (*Anagallis arvensis*) Stem is smooth and hairless. Star-shaped flowers are salmon pink and have five non-cleft petals. Sometimes the flowers are white. Similar growth pattern to chickweed. Best way to tell apart (when not flowering) is to look for the hair strip on the stalk. Scarlet pimpernel does not have a hair strip. **Poisonous.**
- **Common mouse-eared chickweed** (*Cerastium fontanum*) Grows spindlier. Leaves are more rounded. The entire plant is pubescent (has fine hairs). The stalk is terete (round in cross-section) and has hair all around rather than in one strip. The flower has a pistil that is split in five rather than three. Edible.
- **Petty spurge** (*Euphorbia peplus*) Exudes milky white sap when cut. Hairless. Flowers are green. Leaves are teardrop shaped with an entire margin. The plant is Mediterranean in origin but has naturalized in North America. Caution is advised as it can be **poisonous.**
- **Lesser chickweed** (*Stellaria pallida*) Lack the flower petals. At times it may have miniscule petals. It also has a strip of hair down the stalk. Can be used the same way.
- **Greater chickweed** (*Stellaria neglecta*) This plant is sometimes clumped together in the same species as *S. media*. Greater chickweed is a larger, more robust plant. Its petals are the same length or longer than the sepals. It also has the strip of hairs on the stalk. Greater chickweed has a shorter flowering season than *S. media* (which flowers throughout the entire season). *S. neglecta* only flowers in the early to midsummer. Can be used in the same manner.

White Chanterelle

Cantharellus subalbidus | Cantharellaceae

Synonyms

None

Description

Large (5 to 15 cm) vase-shaped white fungi found in conifer forests. Will reappear each year in the same spot if there has been enough rain. This mushroom will always **grow on the forest floor, never on wood.** They often grow alone or scattered, sometimes in small family groups with a big one and a few little ones attached. You will not see large attached groups of multiple white chanterelles. If you find white mushrooms that grow densely clustered, they might be crowded white clitocybe (*Clitocybe dilatata*). These are poisonous. See look-alike section.

Figure 38.1. White chanterelle.

Cap is white, broadly convex to flat when young, and with age, it develops a depression in the middle. Edges of the cap start off rounded and with age become uplifted and wavy. The underside has **false gills**, really just grooves or folds that resemble gills; they also produce **spores**. Chanterelles have false gills that run partway down the stalk (decurrent). They are irregular, often forked, sometimes with a corrugated look. **Stalk (stipe)** is 2 to 5 cm long and 1 to 2.5 cm thick, sometimes even larger. It is solid the whole way through. **Flesh** is white with some bruising in yellow to orange tones. This mushroom is not brittle. It has a faint pleasant smell, kind of fruity. I find that the yellow varieties of chanterelle have more of a scent. Some liken it to apricot. **Spore print** is white.

The Latin name *Cantharellus* means "little drinking cup," and chanterelles are sometimes characterized by their vase shape. *Albidus* means "whitish" and *sub* is a prefix that can mean "partially, almost, rather, becoming, or under."

Picking wild mushrooms

In the Pacific Northwest, you can find Pacific golden chanterelle (*Cantharellus formosus*) and white chanterelle (*C. subalbidus*). In the Okanagan, I find white chanterelles rather than golden. After picking chanterelles a few times, you will feel confident that you can identify them. It is very important when you are foraging for wild food that you have correctly identified what you are about to eat. Never is that more important than when dealing with mushrooms. There is no room for error; some mushrooms can kill you or cause life-long impairments, such as failed kidneys. So please be careful, do not take any chances. The world of mushrooms is a fascinating one, and if you treat it with respect, you can have a lot of fun learning and even eating the safe ones. It is best to have several mushroom guides to refer to. There is also a lot of good (and some confusing) information online. The best way to learn about local mushrooms is to take a course in your area or learn in-person from an experienced mushroom hunter.

Whenever I plan to eat a new mushroom, I make sure that I have done my research. Have I looked at all identifying features? Does the mushroom tick all the boxes of the species I think it is, not just maybe, but certainly? Wishful thinking has no place in mushroom hunting. Have I looked up any possible look-alikes? If some of the look-alikes are deadly, I might skip even trying. Sometimes I observe a mushroom species for a few years before attempting to eat it. I always take a spore print. This is easily done by putting the mushroom cap down on a piece of paper; if the mushroom has white spores, they will not show up very well on white paper, so use colored craft paper to get greater contrast. I place an upside-down cup on top of the mushroom to ensure that the spores do not float away. Leave for a few hours or overnight. Uncover and see what beautiful art the mushroom has created.

You do not need to know every mushroom in the book in order to enjoy eating one, but

Figure 38.2. Note the decurrent false gills.

you do need to **become an expert on the mush room you are about to consume**. You must know if there are any poisonous look-alikes and what their symptoms are.

Once you have properly identified a mushroom, it is time to sample it. Put one intact mushroom aside. Why, you ask? This mushroom will come with you to the hospital if you start to develop symptoms. When testing an unknown mushroom, I make sure to do so within travel distance of the hospital. I will not try a new mushroom if I am out in the back country. Mushrooms that you already know how to identify and know your reaction to are fine to eat in remote areas. Not the new ones.

Whenever I introduce a new mushroom, I eat only a little bit. The theory being, less mushroom, less severe reaction. This is in case I have misidentified or have an allergic reaction. Not everyone reacts the same to a specific

Figure 38.3. Fall chanterelle harvest.

mushroom. For example, I can eat a platter full of golden chanterelles, but they make my grandmother ill. She becomes nauseated and develops diarrhea. The mushroom books cannot know facts like this. Therefore, show restraint, even though the mushroom might taste amazing! Because certain mushrooms can be hard to digest, eat them in smaller quantities as flavor enhancers rather than the main course. Too much of anything can make you feel sick. You do not want to gorge on a mushroom the first time, only to feel sick and never dare eat it again. Another fact, some mushrooms react adversely with alcohol. I try to avoid alcohol in the days around eating a new mushroom just to be certain.

I make a point of being the only one in the family to eat the mushroom. No one else gets to try it until I have tested it without a shadow of a doubt for safety. Some mushrooms will not give adverse reactions until several days later. It is advisable to not eat mushrooms for too many days in a row as they can be hard to digest. With very few exceptions, cook your mushrooms thoroughly. Not only do they taste better, but they are also more easily digested that way. We can access the nutrients better in a cooked mushroom.

Uses

In my opinion, there is nothing as delicious as chanterelles fried in butter and eaten on toast. Before cooking them, I gently clean dirt away with a paint brush. I avoid washing the mushroom unless there is no other way to remove the grit. Washing results in a mushy, less flavorful mushroom. I chop them in smaller pieces. When I cook the chanterelles, I dry-fry them first. They sometimes contain a lot of liquid, and this will come out in the pan. Once that has cooked away, I add a generous serving of butter and some salt and fry until slightly crispy. Mouth-watering! They have a meaty texture and a unique flavor which is why they are sold in specialty stores at premium prices. The cooked mushrooms can be added to sauces, stews, and omelets or any other meal. They will be the star of the show, no matter which meal you add them to. To preserve, I usually fry them until the water is gone, and then I freeze them. I do not add butter until I pull them out of the freezer to use. They last for about a year this way. They are not as delicious thawed as they are fresh, but they still steal the show when added to a meal. Many mushrooms also dry nicely, but I find that the chanterelles do not hold up very well to the drying process. They will not be as tasty as they can be, preserved in this way. I have seen recipes for pickled chanterelles, but I have never tried this.

Figure 38.4. Chanterelles on toast.

Other uses

When I harvest the mushrooms, I pluck them from the ground and then trim the bottom of the stem and other unwanted parts and leave those in the woods tucked in under some moss. Some people say that it is better to cut the stem and leave the bottom of the stem intact in the ground. There is some debate about that. In one study done by the Oregon Mycological Society, they found that there was a slight reduction of the mushrooms in areas where the mush rooms had been cut from the ground rather than plucked entire. It might be because the mushroom is attached to a vast mycelium network, and by cutting, perhaps some bacteria is able to enter the network. Perhaps the plucked mushroom seals the opening to the network effectively. Interestingly, they also were able to determine that harvesting in general did not seem to have a detrimental effect on the mushroom. The mushroom part is just a fruiting body of an entire network that forms mycorrhizal relationships with certain trees.

White chanterelle prefers conifers, especially Douglas-fir (*Pseudotsuga menziesii*) and hemlock (*Tsuga* spp.). The fungi have hairlike strands, called hyphae, and these make up the mycelium network. Around the hyphae, a sheath called an ectomycorrhiza binds with the root tips of living trees. This bonding is beneficial to

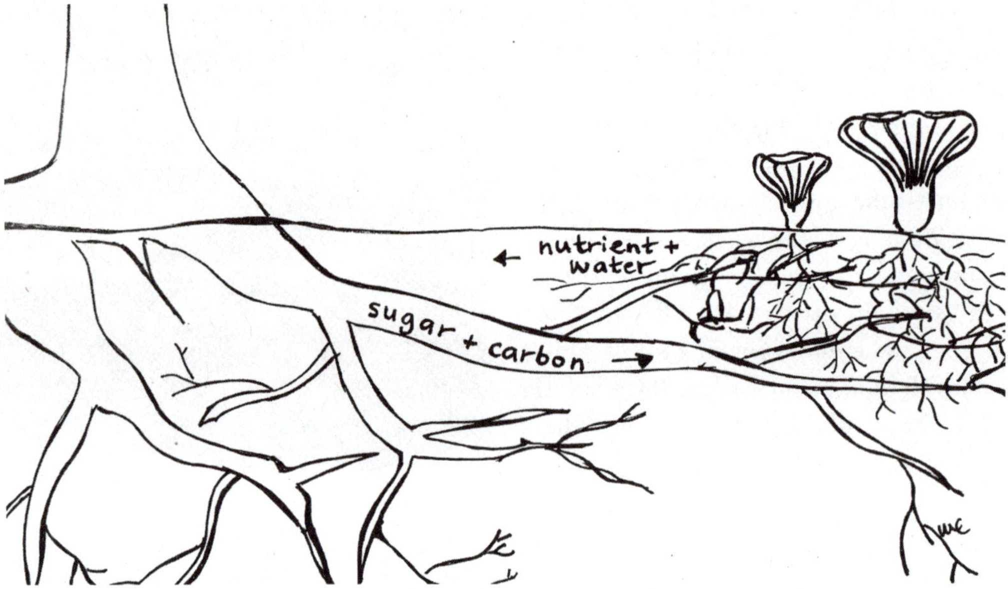

Figure 38.5. Mycelium network connected to trees.

both tree and mushroom; it helps the tree access water and nutrients it would not otherwise be able to reach. The tree in turn repays the favor by supplying the mushroom with carbohydrates and sugars. Chanterelles have long-lasting relationships with their host trees and will not start to produce fruiting bodies until the tree is at least ten years old. Without the forest, the mushroom becomes unable to thrive. Clear-cutting and logging have a detrimental impact on this species.

The chanterelle mushroom will release spores throughout its entire life. I like to pick my mushrooms in baskets with loose weave so that the spores can be spread throughout the forest as I forage. I am hoping that those spores will land in favorable spots and be able to create their own mycorrhizal relationships, and in a decade or so, more yummy fruiting bodies.

Caution

Be sure you have identified the mushroom correctly. See Similar species/look-alikes for more info. The white chanterelle can give some people upset stomach if eaten in too great quantities.

While hunting for mushrooms, keep your bearings in the woods. It is easy to become disoriented and lost if you are looking for mushrooms in an unfamiliar forest. This sounds obvious, but you would be surprised how many times mushroom pickers get themselves lost. Mushroom hunting can become addictive. It might become your new favorite hobby and consume all your waking hours during mushroom season.

Not a lot of research has been done on medicinal properties of the white chanterelle. However, the golden chanterelle (*Cantharellus cibarius*) has been used traditionally in Chinese medicine to prevent night blindness and inflammation of the eyes. It allegedly helps tone the mucus membrane. *C. cibarius* contains eight essential amino acids, vitamins A and D2, and protein.

Similar species / look-alikes

Beware: I might have missed a look-alike that grows in your area. Please use several other mushroom guides as well as this book to properly identify any mushroom you plan to eat.

- **Crowded white clitocybe** (*Clitocybe dilatata*) Large white mushrooms that grow clustered. Has attached true gills that descend the stalk slightly. They are close together, narrow, and whitish. Stalk is dry and fibrous. Spore print is white. Likes to grow alongside roads. I have not seen this mushroom, so I cannot give you first-hand information about it. The information here is from *The Audubon Society Field Guide to North American Mushrooms.* **Poisonous.** Can cause muscarine poisoning. Symptoms: perspiration, salivation, constriction of pupils, blood pressure fall, slowing of pulse, abdominal cramps, diarrhea, tears and blurred vision. Symptoms last for 6 to 24 hours. May require hospitalization. Has caused death. Other *Clitocybe* are also white and can possibly be mistaken for white chanterelle. Please use reputable mushroom guides to learn more.
- **Wood hedgehog mushroom** (*Hydnum repandum*) Looks the same as white chanterelle from above only. Under the cap, it has small spikes rather than gills. Spore print is white. Edible.
- **Fragrant hygrophoropsis** (*Hygrophoropsis olida*) White to pink vase-shaped mushroom. Has true gills descending stalk. These can be forked with cross veins. The gills are thick and close to one another. This species likes the same habitat as the white chanterelle. It is strongly fragrant. Spore print is white. Edibility unknown.
- **Leaflike oyster** (*Hohenbuehelia petaloides*) White to brownish. More fan shaped than funnel shaped. Will have a gelatinous feeling to it rather than the chunky feel of a chanterelle. Grows on logs and stumps. Spore print is white. Edible.
- **Pacific golden chanterelle** (*Cantharellus formosus*) Looks like the white chanterelle but is yellow and often smaller. Edible.

Figure 38.6 Pacific golden chanterelle. Credit: Kate Wamboldt.

Lobster Mushroom

Hypomyces lactifluorum | Hypocreaceae

Synonym

Lobster fungus

Description

Despite looking like a mushroom, lobster mushroom is a **parasitic ascomycete** fungus. It invades other mushrooms and then transforms them into large solid orange structures. Most often its hosts are mushrooms in the genera *Russula* or *Lactarius*. Frequently, the host cannot be identified so caution is advised. See caution box.

Found on forest floors in conifer forests with scattered deciduous trees. The lobster mushroom looks slightly deformed and often forms a **fluted structure with a short stalk.**

Figure 39.1. Lobster mushroom.

Surface of the mushroom is bumpy and often rippled. It can be very large (up to 30 cm across and 25 cm high) with a density that few other mushrooms have. **Flesh** is solid and lighter yellow white on the inside and bright orange on the outside. Has white **spores**; under a large specimen, the ground is often white. Smells a bit like seafood. As it gets older, its color changes to a dark red and its smell increases. Grows solitary, but a thorough search of the nearby area usually unearths a few more. Top is often covered in dirt that can be hard to remove without cutting part of the mushroom off.

The name *Hypomyces* is Greek and means "a mushroom underneath"; *lactifluorum* is Latin and means "flow of milk" and could refer to a host mushroom in the *Lactarius* genus; some which have milky juice when cut.

Uses

I still remember the first time I found a lobster mushroom. It was on Pender Island, BC, and my husband and I were out walking on a beautiful trail. We saw first one of these mushrooms, then another, and another, and another. I did not have my mushroom guides with me (they were still in the tent trailer by the motorcycle), but I remembered something I had read in the text that they were an edible species. We picked some and brought them back to camp with us. The first thing I did when we got back was dig out my mushroom guides and read up further on lobster mushrooms. We fried those mushrooms in butter over our camp stove that night, and it is one of my favorite foraging memories.

So, you might say, what about observing a species cautiously for a while before consuming it? What about eating a new mushroom close to access to medical help? What about only one person consuming a mushroom initially? What about only eating a little the first time? What about making a spore print?

You would be right! I was carried away by the excitement of picking mushrooms. Which is why I tell you this story. It is so easy to get excited when picking mushrooms and to be less cautious than you should be. We were lucky that day. Lucky that we had identified it properly, lucky that the lobster mushroom had not parasitized a poisonous mushroom, lucky that neither one of us had a bad reaction. But it could have been different. Like I say to my children, do as I say, not as I do.

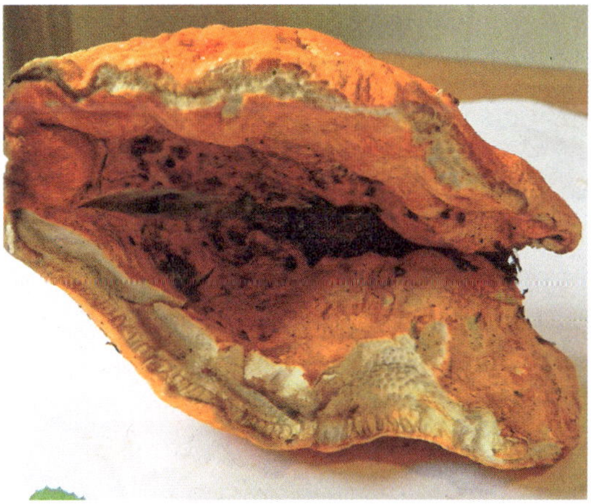

Figure 39.2. Cut edge on older specimen. Note the row of flask-like vessels, called perithecia, where the spores are produced.

Figure 39.3. White interior.

It can be hard to see which mushroom is the host for the lobster mushroom as it engulfs the entire host and leaves no trace of it. The best way to observe it is to become familiar with the area where the lobsters grow and observe it throughout the year and see what other mushrooms you can spot. Here on the farm, there is an area where the main mushroom you see in early fall is short-stalked white russula (*Russula brevipes*). This mushroom is not poisonous; it just does not taste very good...that is, until the lobster mushroom fungus gets a hold of it and transforms it into a delicious treat. Once that happens, you do not see a trace of the short-stalked white russula anymore. In areas you have never been before (like me on Pender Island), it is harder to know what kind of mushroom the lobster has taken over. Therefore, it is better to hunt for this species in areas you know intimately.

I have perused mushroom field guides and other mycological literature for any reports on known poisonings from lobster mushroom, but

I have found none. That does not mean that there is none, but it does mean that it is not common. Lobster mushroom is sold commercially.

Over the years, lobster mushroom has become one of our family's favorite mushrooms. When we find some, there is usually more than we can eat in one sitting, so most often I cook up the entire batch and freeze it in small portions. It is a joy to be able to pull out a package of lobster mushroom and add it to a pasta dish or a paella whenever you feel like it. The mushroom has a sturdy texture, and the red skin contrasts nicely with the pale interior. It both tastes and smells slightly like seafood. I think it is a great replacement for shrimps in a paella. It can also be dried with good result.

Lobster mushrooms contain vitamin D and iron. They also have some calcium and dietary fiber.

Figure 39.4. Phoenix proudly showing her find.

Other uses

The lobster mushroom can be used as a dye, making colors ranging between yellow to purple with pink and orange in between, depending on what else is added. I have not tried this myself…yet!

> **Caution**
>
> Since lobster mushroom is a parasite on other mushrooms, it could technically take over a poisonous mushroom and possibly become toxic itself.
>
> Some people experience bloating and gas after eating lobster mushrooms. Try a little the first time you sample it.
>
> Do not eat raw; mushrooms contain chitin (the same stuff that lobster shells are made of), and this compound is hard to digest unless cooked. You will not access the nutritional benefits of the mushroom raw since they will be locked away from you within the chitin. Once cooked, the nutrients in the mushroom become available to human digestive systems.

Figure 39.5. Peppers stuffed with lobster mushroom and rice.

Similar species / look-alikes

- **Cinnabar-red polypore** (*Pycnoporus cinnabarinus*) This mushroom has a bright orange color like the lobster, but once you turn it over, you will see that it grows out of dead wood rather than on the ground. It has spongy pores on the underside. Not edible. Has some medicinal properties.
- **Beefsteak polyporc** (*Fistulina hepatica*) Fleshy and squishy mushroom that is orange red. It grows on dead oak. Spore print is pink/orange. Edible. The information here is taken from *The Audubon Society Field Guide to North American Mushrooms* since I am unfamiliar with this species.
- **Scaly vase chanterelle** (*Gomphus floccosus*) Orange, yellow, or reddish mushroom (5 to 15 cm). Vase shaped with scales both on top and inside of vase. Outside surface is ridged down to forest floor. Spore print is ochre (brown/orange). Edible, but not recommended due to an indigestible acid.

Yarrow

Achillea millefolium | Asteraceae

Figure 40.1. Yarrow in September.

Synonyms

Bloodwort, achillea, carpenter's weed, wound wort, devil's nettle, old man's pepper, thousand-leaf, staunchweed, nosebleed plant, thousand-seal, herbe militaris, gordaldo, sanguinary, milfoil, soldier's woundwort, plumajillo, yarro-way, knight's milfoil, field hop, arrowroot, death flower, eerie, hundred-leaved grass, knyghten, old man's mustard, seven-years love

Description

Perennial herb (20 to 60 cm tall). **Stem** is hollow, ribbed, and covered in cobwebby white hairs. **Leaves** are fern-like, dark green, and elliptic, widest in the middle. They often have white cobwebby hairs. Each leaf has numerous leaflets. Leaflets vary between simple-pinnate or double-pinnate form. See drawing on page 150. Leaves have a strong medicinal smell when crushed. **Flowers** are held in compound corymbs. Each flower has five white ray petals, sometimes notched at the tip, surrounding five or more white disk florets. In each disk floret there is a yellow style. The **achenes/seeds** are flattened and oblong. These dry on the corymb and turn brown. **Root** is a spreading rhizome.

The name *Achillea millefolium* means "Achilles' thousand-leaved plant." According to Greek mythology, Achilles was dipped in yarrow water and became invincible. The only part of him that could be injured was the heel

Figure 40.2. Flower.

Figure 40.3. Seeds in October.

that was not dipped. Achilles henceforth used yarrow to heal his soldiers on the battlefield in the Trojan War. Many of the common names of yarrow refer to its use by soldiers.

Uses

At this time of year, there are fewer greens to pick. Yarrow is still looking fresh and green though. I like nibbling on the leaves as I go for walks, and I usually pick a little to bring home to make fresh yarrow tea. The flavor is strong with aromatic, bitter, and sweet all mixed together. It makes an excellent herb to add to a stew or a bake. Make sure you do not add too much, or the other flavors will not come through. Fresh yarrow can be used sparingly in a salad too. The texture of the leaves can be hard for some people to get used to: it is kind of rough and hairy. If this bothers you, the leaves can be cooked.

I add dried crumbled yarrow leaves to a salt grinder with large sea salt and other herbs and voilà, wild herb salt!

Figure 40.4. Wild herb salt.

It is not only the leaves of yarrow that can be gathered at this time of year. The dried flower heads with seeds and husks can be picked and stored in a dry place. They smell amazing, and I like to use them in tea, as spice, or as an ingredient in yarrow lozenges.

Earlier in the year, the flowers can be used in tea or salads. I like adding a few flowers to bread dough to make it taste more interesting.

In Scandinavia, yarrow was historically used in beer brewing. Analyses of old Viking settlements allude to yarrow as a plant that they cultivated and/or encouraged around their dwellings and used for food, drink, and medicine.

Both flowers and leaves are high in nutrients and phytonutrients such as flavonoids, carotenoids, and terpenes.

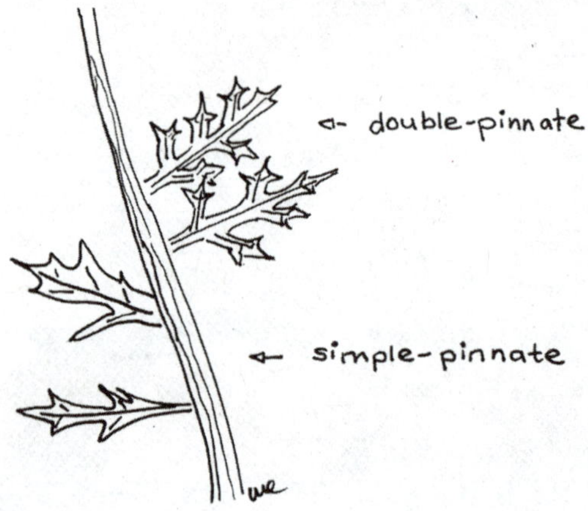

Figure 40.5. Comparison between simple-pinnate and double-pinnate leaf. Yarrow can have both.

Other uses

Yarrow is one of the oldest known plants to have been used by humans. It was found buried together with a *Homo neanderthalensis* in Shanidar Cave, Iraq. The grave was dated to 65,000 BP (before present).

Yarrow is a styptic. The plant contains a chemical called achilleine, known to stop bleeding. When used for this, it is dried, then ground to a fine powder and sprinkled over the cut. It can be used to stop a nosebleed. Use either a fresh leaf or dry powder.

The plant has natural antiseptic qualities and can prevent a wound from getting infected. It also contains some salicylic acid and can help reduce inflammation and lessen pain. The plant is best harvested for medicine when it is in full bloom. Both leaves and flowers are used.

As a tea, it can be helpful against menstrual cramps and for a regular menstrual cycle. This plant should be avoided during pregnancy. See caution box.

I like using yarrow in tea to combat flu symptoms. I mix it with peppermint and elder flowers; this potent brew helps to raise your body heat to kill the pathogens, as well as soothes your throat and boosts your immune system. I also find yarrow tea helpful for keeping my fingertips warm; the herb reputedly increases blood circulation.

Chewing on a yarrow plant (any part) can relieve pain in the mouth. I have used it to lessen pain from canker sores, and I have read about people using it this way for toothache. We like to use it as a mosquito repellent: either burn it fresh over your campfire to keep you safe from the mosquitoes or rub it directly onto your skin (some people might react adversely to this).

Yarrow is an excellent companion plant in the garden. It attracts ladybugs and predatory wasps that help to keep pests in check. When put in the compost, it helps to speed up decomposing and bacterial activity.

Yarrow Throat Lozenges

¾ cup yarrow infused water
1 cup sugar
½ cup honey

- Boil water, pour over dried yarrow seeds (2 to 3 tablespoons). Let steep for 20 to 30 minutes, covered to contain the beneficial volatile oils.
- Strain the seeds out and use the infused water to make the lozenges.
- Mix water, honey, and sugar well, keep stirring over medium to low heat until it hardens somewhat and becomes fluffy. This will take forever (20 to 30 minutes or so). Keep stirring the bottom of the pot and avoid the sides. Make sure that it does not burn. When the mixture reaches a high enough temperature (280° to 300°F, 140° to 150°C), it is ready to be poured into molds or onto a tray. Silicone molds and silicone trays work the best but regular metal baking sheets work too.
- If you do not have a candy thermometer, drop some of the mixture in a cold glass of water; it is ready when it forms a hard ball.
- Let lozenges dry for several hours and toss with icing sugar if you wish once they are hard. Store in a dry place.

Figure 40.6. Yarrow seed throat lozenges.

Caution

Use over long periods of time can cause the skin to become more light-sensitive. People with allergies to plants in the aster family can develop skin rashes from the yarrow. Not to be used during pregnancy; it can cause miscarriage. Stop using at least two weeks before scheduled surgery, it may increase risk of bleeding. Consult with your doctor before using if you are taking blood pressure medication or anticoagulants. Make sure you have properly identified the plant so as not to mistake it for poison hemlock, see Similar species/look-alikes.

Figure 40.7. Tansy.

Figure 40.8. Birchleaf spirea.

Figure 40.9. Pineapple weed.

Similar species / look-alikes

- **Tansy** (*Tanacetum vulgare*) Larger plant (50 to 150 cm) with yellow button-like flowers in flat-topped clusters. The finely divided leaves are aromatic. Use with caution.
- **Pineapple weed** (*Matricaria discoidea*) Likes to grow by roadsides (5 to 40 cm tall). Leaves are finely dissected and aromatic. Flowers lack ray petals and smell like pineapple when crushed. They look like a daisy that has lost its outer petals. Edible.
- **Birchleaf spirea** (*Spiraea betulifolia*) The best way to tell them apart are by the leaves. Leaves resemble birch leaves; they are egg shaped with jagged edges, alternate on stalk. Flowers are creamy white and fuzzy, borne in dense flat-topped clusters. Each flower has five petals and multiple stamenss. Not aromatic. Has medicinal qualities.
- **Douglas' dustymaiden/hoary false yarrow** (*Chaenactis douglasii*) Likes to grow on open, dry, and rocky areas at higher eleva-

Figure 40.10. Poison hemlock leaves — deadly!

Figure 40.11. Poison hemlock flowers. This plant can kill you!

tion. Leaves are similar to yarrow, but they are more silver toned and hairier. Flowers consist of clusters of five-petaled tubular flowers with pink/purple styles. Used medicinally.

- **Poison hemlock** (*Conium maculatum*) Does not bear much resemblance when mature but could possibly be mistaken for one another when very young. Leaves are divided two to four pinnately. They are triangular rather than linear. They do not have the aromatic smell of yarrow. The plant resembles a carrot and is in the Apiaceae/Umbelliferae family. The five-petaled white flowers are borne in clusters. It lacks the disk florets that yarrow has. Stem often has purple blotches on it. Grows much taller than yarrow (up to 1.5 m). **Deadly poisonous!**

Rosy Gomphidius

Gomphidius subroseus | Gomphidiaceae

Synonyms

Rosy slime spike

Description

Gilled mushroom that likes to grow in areas with Douglas-fir and other conifers. It grows alone or scattered **on the forest floor. Cap** can be 3 to 6 cm wide and is covered in a **slimy veil.** The color of the cap can range from pink to red to olive. When it matures, it can sometimes develop darker spots. The shape of the cap is rounded when young, convex at maturity. **Gills** are decurrent; they run down the stem somewhat. Gills are close to one another, and short gills on the edge of the cap are frequent. The gills are pale white/grey initially and later become smoky grey. **Stem** is 3 to 9 cm long and 0.5 to 1.5 cm thick. It commonly tapers toward the bottom. Stem is white at the top and bright yellow at the base. It bruises black. Older mushrooms often have a hairy, poorly developed ring. **Ring** becomes darker with age as spores land on it. **Flesh** of the mushroom cap is white, and the flesh of the stem is yellowish. **Spore print** is brown/black. This mushroom does not have a distinct scent.

Interestingly, this mushroom is often found growing near Lake's boletus (*Suillus lakei*). Re-

Figure 41.1. Rosy gomphidius. Note the yellow tapered base.

search has found that species of *Gomphidius* parasitize *Suillus*. The *gomphidius* attaches to the mycorrhizae created by the *suillus* and diverts some of the nutrients to itself.

The genus name comes from *Gomphos*, an old Greek word that means "large wedge-shaped nail" or "plug." The *sub* in *subroscus* is to differentiate it from *G. roseus*, European rosy *gomphidius*. *Roseus* means "rosy."

Uses

Gomphidius is a fairly new mushroom for me. I have only been eating it for the last five years. There are quite a few different gomphidius, and they can be hard to tell apart. To be perfectly honest, sometimes I do not know if what I have is a slimy gomphidius or a rosy gomphidius. Their cap colors overlap somewhat. Here on the farm, we seem to have more rosy gomphidius, but some of them look suspiciously like slimy gomphidius because of their darker cap in brown/wine colors. Luckily it does not matter which one of them you have since they are both edible. I think that they have a very similar taste. See similar species for more info on how to tell them apart. I use the telltale tapered yellow-based stem as the main clue to identify a gomphidius from other types of mushrooms.

Right now, the woods are teeming with Lake's boletus (*Suillus lakei*) and blue-staining suillus (*S. caerulescens*). They are not the tastiest of mushrooms, so I keep searching around them and often I am rewarded with a few rosy gomphidius. I like picking them when they are small, before their cap has flattened out. They still taste nice when mature, but they look so aesthetically pleasing when they are little. They really do look like wedge-shaped nails.

They are not often attacked by bugs. However, they tend to have forest debris clinging to their sticky cap and can look quite dirty. As you peel off the slimy veil, all the dirt comes off in one easy swoop. The veil should be removed before cooking; it can cause digestive upset if eaten. I usually fry the mushrooms before adding them to any dish. Initially I dry-fry them

Figure 41.2. Rosy gomphidius slimy veil. Remove before eating.

Figure 41.3. Spore print.

Figure 41.4. Note faint ring and short gills on edge of cap.

until all the liquid is gone and then add butter and salt. Their flesh changes from white to pink when cooked. Their flavor is mild and pleasant. I like adding them to stir-fries, casseroles, and sauces rather than eat them on their own. The flesh of the mushroom, when cooked, is soft and chewy. Rosy gomphidius are lovely paired with Chinese black bean sauce or added to a ginger beef stew.

If I have great quantities of the mushroom and need to preserve some for the future, I cook and freeze them. They blacken over time in the freezer and when dried.

Be cautious of where you pick this mushroom as it can accumulate heavy metals. See other uses and caution box.

Other uses

The related slimy gomphidius (*Gomphidius glutinosus*) has been used as a bioremediation tool. As a hyperaccumulating fungus, it can absorb and concentrate elements such as radioactive cesium-137 more than 10,000 times the background levels. Fungi hold great promise as a way to clean up pollution and contaminated sites. The related *G. viscidus* is used in China as a medicine for neurodermatitis.

Caution

This mushroom can absorb heavy metals and toxins in high quantities. Avoid picking it near industrial areas, gas stations, or heavily trafficked roads. Peel off the slimy veil before consuming; the veil can cause digestive upset if ingested.

Similar species / look-alikes

- **Rosy gomphidius** (*Gomphidius roseus*) European species with the same common name and almost identical Latin name. *G. roseus* (European) has a brighter pink cap and does not have the yellow base on the stem. Edible.
- **Slimy gomphidius** (*Gomphidius glutinosus*) Can be very hard to tell apart from rosy gomphidius. Cap tends to be more purple or brown. Base of the stem is bright yellow. Stem tends to be thicker than rosy gomphidius. Grows alone or scattered. Edible.
- **Clustered gomphidius** (*Gomphidius oregonensis*) Prefers to grow in clusters with the bases attached to one another. Cap is salmon to brown colored, and it often has black spots. Can look quite unappetizing. Stem has yellow base. Edible.
- **Fragile russula** (*Russula fragilis*) Cap can be pink, wine, olive-brown, green, or yellowish. It does not have a slimy veil but can be

slimy and sticky when wet. Gills do not go down the stem at all. Stalk is bright white and does not have a ring. The entire mushroom is very brittle. Spore print is yellow. **Not edible**.

Figure 41.5. Fragile russula on left, rosy gomphidius on right.

There are many similar-looking russulas, and even mushroom experts have a hard time identifying the species with certainty. They are easy to tell apart from the *Gomphidius* though as all *Russulas* have gills that do not go down the stem. They also lack the yellow tapered base of the stem and the slimy veil.

- **Short-stalked suillus** (*Suillus brevipes*) Has pores rather than gills. Has a sticky coating on cap that needs to be removed before eating. Cap is reddish brown. Does not have a ring on stem. Stem is bulbous at bottom. Spore print is cinnamon colored. All the *Suillus* tend to be invaded by grubs. Edible.
- **Pine-spike** (*Chroogomphus* spp.) Was formerly part of *Gomphidius* but have been moved to their own genus. I am not personally familiar with these mushrooms, so the information is taken from *The Audubon Society Field Guide to North American Mushrooms*. These mushrooms have gills that descend the stalk, like the *Gomphidius*, and have a hairy ring on stalk. They do not share the slimy veil and the yellow base of the stem. Color of the flesh is pinkish rather than white. Spore print is grey/black. All *Chroogomphus* mushrooms are believed to be edible, but I advise you to look up individual species in detail in reputable mushroom guides before attempting to taste them.
- **Slippery Jack** (*Suillus luteus*) Can only be confused with one another until you turn it over and see under the cap. Slippery Jack has spongy pores under the cap rather than gills. This mushroom also has a slimy veil that needs to be removed before consuming. Top of cap can be red, yellow, or brown. Stalk often has a ring. Spore print is cinnamon. Edible.

Figure 41.6. Short-stalked suillus.

Black Walnut

Juglans nigra | Juglandaceae

Synonyms

Eastern American black walnut, American walnut, walnut

Description

Stately deciduous tree that can reach 30 m. More common in eastern areas of Canada and the US. Has a deep tap **root** with several supporting lateral roots. **Bark** is grey brown and deeply grooved. **Leaves** are compound, and each has 15 to 23 leaflets, attached in pairs. Each leaflet is lance shaped, has a serrated edge, and the tip tapers to a point. Leaves are glossy, dark green above, and paler and hairy below. The end leaflet is often missing. **Stems** are fuzzy and sturdy. The fastening point of the petiole, the **leaf scar**, has a distinct heart shape. It has three darker markings; to me the leaf scar looks like a monkey face with two eyes and a smiling mouth.

Figure 42.1. Furrowed bark on older black walnut tree.

The **flowers** appear at the same time as the leaves. Black walnut is monoecious (both male and female flowers are on the same tree). Male flower is a long green catkin; female flower is a shorter yellowish frilly spike. At the bottom, it has a green bulb that will swell upon pollination and become the **fruit** (drupe). They tend to form in pairs at the end of stems but can also form in clusters of three or be solitary. Fruit is fleshy, green, and smells strongly aromatic. As it matures, it becomes more yellow and later turns dark brown and mushy. It falls to the ground when mature (October).

Both the green flesh and the overripe fruit stain hands and anything they touch. Inside the flesh is a hard shell that holds the nut. **Shell** is deeply grooved, and if left within the decomposing fruit for any length of time, it will take on the color (dark brown/almost black) of the decomposing fruit. The shell is slightly ovoid; it is incredibly hard. Inside is one (divided in half) drupaceous **nut**, light brown on the outside and creamy white on the inside.

The name *Juglans* comes from Roman times and has its roots in their god Jupiter, who they believed dined on walnuts while residing on Earth. *Nigra* means "black."

Uses

Black walnut has been introduced in the Okanagan Valley, and often you can find them growing in parks or public areas. In October, the ground underneath a black walnut is usually littered with black round fruit that stain the pavement. The decomposing flesh of the walnut can be very unsightly; it is strong smelling and slippery. I bring large plastic bags and rubber gloves for picking. The first year, I made the mistake of picking them with my bare hands. I had stained hands for two weeks no matter how hard I scrubbed them. As soon as I have picked the nuts, I remove the flesh. If you leave the flesh too long around the nutshell, the flavor of the nut will become bitter.

Do not put your walnut waste in the compost; it will stop your vegetables from growing. The walnut tree and all its parts are allelopathic; it sends out a chemical, juglone, that inhibits growth of other plants. It helps the walnut to ward off competition for resources.

Once the flesh is gone, I scrub the nuts. You want to do this carefully because any splatter will stain your walls and counter. I then dry them on paper towels and set them in a cool, dry, and well-ventilated place to cure, turning them occasionally to make sure that all sides dry. The curing process allows the nutmeat to dry, and the flavor changes from sour/astringent to sweet and oily. It takes around a month. Crack open a nut after a month and check it. If it is not sweet yet, leave the rest to cure longer.

Figure 42.2. One leaf. Note the lack of leaflet on tip.

Figure 42.3. Immature fruit in July.

Figure 42.4. Close-up of fuzzy leaf stalk.

Figure 42.5. Leaf scars look like smiling monkey faces.

Figure 42.6. Cross section of black walnut with nutmeat inside.

Cracking open black walnut shells is a feat. Your regular nutcracker does not stand a chance; ours broke in two when we tried it. The first year, we even tried running them over with our car to crack them, and it did nothing to the shells. They are possible to smash with a hammer, but this method seems unsafe as the shells ricochet in all directions. The shells are so hard that they have been used in oil well drilling as well as in metal polishing. Nothing I want flying around my face. The best way to open them, we found, was to use a vise grip tool. Even when you crack the shell open, it is still hard to get the nut out intact due to a lot of small grooves and crevices inside the shell.

The flavor is nothing like an English walnut (*Juglans regia*); instead it has a rich, almost liqueur-like flavor. The nuts can be used as you would English walnut; we eat ours as a treat raw while cracking them. They taste divine lightly roasted too. In our family. there is never enough to add to baking. The nuts contain omega-3, magnesium, protein, dietary fiber, zinc, iron, copper, potassium, selenium, vitamins E and B6. The nuts are beneficial for both brain and skin.

The nuts store for about a year in a cool and dark place if still in shell. I have had some for two or even three years that have still been good, but as with all oily nuts, they can turn rancid with age. If the nut is brown or has an inner yellow core after cracking open, discard it. Shelled, their shelf life is shorter. To keep shelled nuts for longer, keep them in the fridge or freezer.

Some people pickle the green fruit. A good friend of mine gave me some, and they are fantastic with crackers and cheese! Another use for the unripe fruits is to make nocino, a sweet liqueur. You want to pick them before the shell inside has become hard. Combine unripe walnut fruits, spices, and alcohol and let sit for forty days. Later in the process, syrup is added, and additional time is allowed for the flavors to combine.

Figure 42.7. Dye bath from ripe walnut fruit husks.

Other uses

Black walnut is often part of herbal remedies for parasites. The green fruit or leaves can be used to make a tincture for easy all-year use. This can be added to a drink or taken in a dropper. It works for both humans and animals with varying results. I like adding a dropper-full of the tincture to a big glass of water, with some cinnamon and ground cloves. You can add lemon juice too. This concoction taken on an empty stomach in the morning is my go-to parasite purge. I take it on a regular basis just to make sure that my body is inhospitable to parasites. Living with pets and working with soil, there are lots of opportunities for parasites to be ingested. I do not use this for more than two weeks at a time before taking a break. The tincture should not be used by breastfeeding mothers. See caution box.

Caution

The tincture of leaf or unripe fruit should not be used if pregnant or breastfeeding. It can stop lactation. The active chemicals in the black walnut hulls or leaf can be transferred in the breast milk to the baby and are not safe for an infant or fetus. It can also have a laxative effect.

While using a parasite purge, you may feel nauseated and dizzy. If that happens, lower the dosage or take a short break from consuming it, then gradually increase the dosage again to a level that your body can handle.

The walnut tree and all its parts have allelopathic properties and will hinder growth of other plants in the area around it. Consider carefully where to plant a walnut.

Use gloves when picking, the walnut flesh stains everything!

Black walnut wood is sought after by cabinet makers because of its beautiful rich red/brown color and decorative grain. It also has a natural resistance against decay and insects. The wood is hard but can be worked easily with both hand and machine tools.

The hulls of the walnut (once they have turned mushy brown) can be used as a dye to give a solid brown color. I dyed some cotton cloth with walnut husks, and the color attached to the fabric without any need for a mordant (fixative) and stayed through machine washes. I found it to be a dull kind of brown.

Figure 42.8. Leaf of English walnut.

Figure 42.9. English walnuts on left, black walnut on right.

Similar species / look-alikes

- **English walnut** (*Juglans regia*) Leaves have fewer leaflets (5 to 9), and the end leaf is large and oval. Fruit splits open cleanly when ripe and sometimes exposes the nutshell while still on the tree. Husk of the nut is not as grooved as black walnut. It is much easier to crack. Edible.
- **Heartnut** (*Juglans cordiformis*) Native to Japan but has been introduced as an ornamental. Leaves have 11 to 17 leaflets. Fruit grows in clusters of 4 to 10. Fruit is green and oval. Both the shell and nut are heart shaped. Edible.
- **Butternut** (*Juglans cinerea*) Native to eastern North America. Leaves have 11 to 17 leaflets. Fruit grows in clusters of 2 to 6. Fruit is oval; the nutshell is oval and deeply furrowed. Hard to crack. Edible.

Chicory

Cichorium intybus | Asteraceae

Synonyms

Blue dandelion, succory, blue weed, blue sailors, coffeeweed, cornflower, hendibeh, horseweed, ragged sailors, wild endive, blue daisy, wild bachelor's buttons, radicchio, witloof

Description

Annual or biennial herb (up to 1.5 m). Often flanks the roads or grows in heavy clay soil. Can handle drought well. **Stems** are fibrous and gently ribbed; lower stems are often hairy. Stems are green, with a purple base. **Leaves** are alternate and clasp the stem, becoming smaller the further up the plant you go. They are lanceolate with jagged teeth around the margin, like dandelion leaves. Chicory has fine hairs on the bottom midrib of leaf. Lower leaves form a basal rosette. When cut it exudes a milk-white latex. **Flowers** (2.5 to 4 cm across) are sky blue and have 10 to 20 ray florets. Each ray floret has five teeth on the tip. As it ages, the flower fades to white. In the middle of the flower, blue stamens are held around the bifurcated style, giving the center a curled decorative look. Flower closes overnight or on overcast days. Chicory **seeds** are flattened, ribbed achenes. **Root** is a stout fleshy taproot.

Cichorium comes from *kichore*, the ancient Greek common name for chicory. The name *intybus* comes from the Egyptian *tybi*,meaning "January," the month that chicory was customarily eaten there.

Uses

In late fall, I pick our yearly supply of chicory root. The roots will be bigger as the plant has redirected all its nutrients back into the taproot for storage over winter. The trick is to dig it up before the ground freezes. To ensure that there will be a healthy stand of chicory for years to come, I replant any little root I might unearth, as well as plant some seeds in the disturbed soil.

Before eating, I give the root a gentle scrub under water and sometimes remove the inner core if it is too woody. The flesh of the root can be eaten raw, if you appreciate bitter flavors, or it can be added to cooked meals. I like mixing it with potatoes, Jerusalem artichoke, burdock, and sweet potato in bakes and soups.

Another excellent way to use it is to dry tiny chunks and then roast them. The smell when it

Figure 43.1. Flower.

Figure 43.2. Flower bud.

Figure 43.3. Chicory chocolate bar.

is roasting is divine! I like it well roasted (about 20 to 30 minutes on 300°F [150°C]), to the point where it is quite darkened and crumbles easily in my hand. I eat the roasted chunks as candy or grind them into a powder to use as a coffee substitute. I like it mixed with some regular coffee and with lots of steamed milk. Chicory has been used all over the world as a coffee substitute, and there are some excellent chicory coffee mixtures to be found at the stores. The roasted chicory root powder can be used in a chocolate bar.

Not only is the root excellent food, but the leaves are also fabulous. At this time of year, you can find second growth of fresh leaves in the basal rosette. That they might have been kissed by frost only enhances them and makes them a little bit sweeter. They taste great eaten fresh or added to cooked greens. I like to gently wilt/fry them in a pan with coconut oil, avocado oil, or lard. Add salt to taste and perhaps a gentle sprinkling of parmesan cheese, and you have a marvelous side dish that feels wholesome as well as decadent. The leaves can be eaten throughout the year, depending on your climate. Here in the Okanagan, they are hard to find in midwinter when the snow lies thick, but as soon as the snow melts, they poke up. I have been able to pick them until December in some years.

The endives that we see in the grocery stores are a species of *Cichorium*, and they have often been blanched by being grown in the dark. This makes the leaves white and sweeter. You can do the same with wild chicory. If you cover

Chicory Chocolate Bar

½ cup roasted chicory root powder
3 cups coconut oil
1¼ cups dark chocolate chips
1 cup puffed quinoa or puffed rice
¼ cup finely chopped hazelnuts or walnuts

- Melt chocolate in double boiler, stir in all other ingredients until well mixed.
- Spread on wax paper and put in fridge to cool. Cut or break into squares.
- Melts if temperature is too high; keep in cool pantry or fridge until consumed.
- Tastes great with some chicory/burdock coffee.

the plant as it grows, the leaves will be lighter. I personally like the bitter flavors of the leaf, so I rarely bother with blanching.

The flowers can be eaten during late summer. As soon as they are picked, they start to lose their extraordinary sky-blue color and take on the color of wrung-out old dishcloths, so pick them just before serving. While still blue and radiant, they look great in a salad or sprinkled over ice cream or yogurt. I like to pull the ray petals off the flower structure before eating them as the other flower parts can be hard to chew.

Other uses

The whole plant is filled with beneficial nutrients. It has vitamins A, B, C, K, and flavonoids, as well as iron, potassium, and calcium. The root is particularly good for the liver as it helps the body cleanse itself more effectively. Chicory root is a source of fructose and maltol, and it also contains inulin. The root is grown commercially. Chicory root lowers blood sugar and alleviates constipation. The bitter leaves are helpful for production of bile; a healthy bile flow improves digestion. Chicory leaves can be beneficial if eaten when you feel bloated and have had too many sweets. It helps to rebalance the body and can lessen cravings. I find that if I eat a side dish of bitter greens such as chicory, dandelion, yarrow, or wild mustards with dinner, I am less likely to reach for those cookies for my evening snack.

Chicory has a long history of being used as a medicinal herb. The ancient Greeks and Egyptians used chicory as a liver tonic, as a sedative, and to improve appetite. Syrup of chicory was used as a safe laxative for children. Some modern herbalists use it to reduce cholesterol levels.

Caution

Do not use if pregnant as chicory can increase or hasten menstruation. Can cause diarrhea if eaten in large quantities. The inulin in the root can cause flatulence. People with known gallbladder problems should consult with their doctor before using.

Figure 43.4. Chicory leaf at top with fine hairs on underside, smooth dandelion leaf on bottom.

Figure 43.5. Basal rosette.

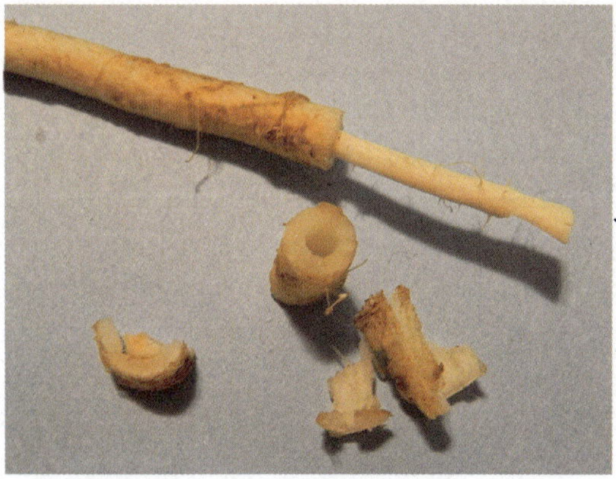

Figure 43.6. Root. Note the inner core.

Figure 43.7. Spotted knapweed.

Similar species / look-alikes

- **Dandelion** (*Taraxacum officinale*) Leaves are very similar. Dandelion leaves usually do not have hair. Leaves grow in a basal rosette; none are attached to the flower stalk. Flower stalk is hollow and not fibrous. Flower is yellow rather than blue. Edible. See dandelion chapter on page 41.
- **Prickly lettuce** (*Lactuca serriola*) Leaves have barbs on the bottom midrib rather than hairs. Flowers are yellow and smaller. Edible. See pricky lettuce chapter on page 49.
- **Sow thistle** (*Sonchus* spp.) Leaves have spiky edges and are glossier than chicory. Flowers are yellow rather than blue. Edible. See pictures on pages 44 and 52.
- **Showy blue lettuce** (*Mulgedium pulchellum*) Formerly *Lactuca pulchella*. The pale blue flowers are similar to chicory, but they are much smaller (1 to 3 cm across). Flowers are spread out on the plant, whereas chicory has its flowers closer along the stem in an almost spike-like fashion. Plant is smaller than chicory. Leaves are lanceolate and have few if any jagged edges. Achenes/seeds have a tuft of hair which chicory lacks. It is a frail-looking plant that is not very common. Edible.
- **Spotted knapweed** (*Centaurea maculosa*) Flowers are more purple pink than blue. Ray florets are skinny. It has overlapping floral bracts with dark edges. Leaves are lacy and deeply lobed. Flowers can be eaten according to some sources; I have not tried this yet.

Dog Rose

Rosa canina | Rosaceae

Synonyms

Dogberry, witches' brier, eglantine, brier, hagrose

Description

Deciduous shrub (1 to 5 m). Introduced from Europe, northwest Africa, and western Asia. Grows in open brushland or waste places; can be quite invasive. Forms **root runners** and expands its territory this way. Can handle some shade and is not troubled by frost nor heat. **Stems** are sturdy and have sharp **hooked prickles**. To me the prickles look like shark fins or perfectly formed waves. **Leaves** are divided into 5 to 7 leaflets that are oval with jagged edges. Clasping the base of the leaf stalk/petiole are two stipules that have grown into one leaf, giving it a butterfly look. **Flower** (4 to 6 cm across) is most often pale pink but can vary between white or dark pink. It has five petals, five green sepals, and numerous yellow/orange anthers surrounding the yellow-green stigma at the center. Petals are heart shaped with delicate veins and wavy edges. Has an oval **fruit** (rosehip) (1.5 to 2 cm long) with five dried sepals forming a star shape at its tip. Hip is orange/red and contains many seeds and **fine white hairs. Seeds** are light-colored.

The reason it is called dog rose is debated. One theory is that using "dog" in the name was a way to say that something was inferior or common. Another is that the root of this plant was used in ancient Greece to cure bites from mad rabid dogs. A third is that the name refers to the prickles that resemble dogs' canine teeth. *Canina* is Latin and refers to dogs. The genus name *Rosa* refers to "rose."

Uses

After frost, the rosehips soften and sweeten. Many types of roses grow in this region; all have hips and can be used interchangeably, but none taste as great as the dog rose, in my opinion. Its flesh is the right combination between sour and

Figure 44.1. Dog rose flower in June.

Figure 44.2. Leaf. Note the fused stipule on base of stem.

Figure 44.3. Prickles: sharp thorns forming from the epidermis/bark on rose branches

Rosehip Soup/Nyponsoppa

12 cups washed rosehips (seeds still inside)
8 cups water
sugar, optional
lemon juice, optional
potato flour

- Combine hips and water and cook down until they have softened and have infused their flavor in the water (15 to 20 minutes).
- Strain through fine mesh to avoid ingesting any of the small hairs.
- Sometimes I pour additional water over the used hips and repeat the process once or twice. The more extra water you add, the runnier your soup will be.
- Add sugar to taste and lemon juice if you want it to be sourer. I often find that neither sugar nor lemon juice is needed; the flavor of the rosehips is perfect.
- I use a little bit of potato flour to thicken it. I mix it in a small bowl with some of the rosehip soup until there are no lumps, then add it back to the main batch, stirring to ensure that no lumps form. You decide how much liquid you want.

sweet, and it has a fair amount on each hip. I am originally from Sweden, and there we regularly consume *nyponsoppa*, rosehip soup. Every grocery store sells it. This soup/drink can be served warm or cold, and it is a great afternoon snack. *Nyponsoppa* made from dog rose is lovely and brings back memories from my childhood.

When using rosehips, make sure that you do not consume the white fine hairs inside them. I was told as a child that they would itch on the way in and on the way out. Here is a recipe for rosehip soup that is easy to make and requires minimal work preparing the rosehips.

I also like to dry a bunch of hips to use for tea. I throw five or six hips in a cup and let them steep for 15 minutes. Strain through fine cloth

Figure 44.4. Rosehip soup.

Figure 44.5. Rosehip. Note hairs and seeds inside.

Figure 44.6. Rosehips in September.

to keep any of the fine hairs out of your drink. The hips are great added to other tea mixtures for additional sweetness and a big dose of vitamin C. They often stay on the bush throughout winter, and although their color changes, they still contain an abundance of vitamin C. In early spring, I have sometimes found rosehips that have become like fruit leather on the bushes, sweeter and less sour. One year my husband made wine from the dried overwintered rosehips, and it was amazing! Rosehip marmalade is another delicacy from this fruit. You have to be careful to scoop out each hip and remove the seeds and hairs before making your jam or marmalade. This often feels too time-consuming for me, but when I take the trouble to do it, I always love the end product.

In early summer, the flowers decorate the bush with their delicate petals and lovely scent. They do not stay long, and if you do not pay attention, you can miss the entire event. Collecting rose petals is a beautiful way to spend a few hours. It requires concentration due to the formidable prickles, which once they have grabbed you, demand a moment of attention as you break free. I like to pick a few petals from each rose rather than whole roses; this way the bees can still find the flowers to pollinate them and a harvest of rosehips later in the year is not jeopardized. There are many things you can do with rose petals. One lovely way is to put them in a jar with water by a sunny window and let them solar infuse for a day or two. I strain the petals out and use the water to drink or to add to face washes or face creams. It has the most delicate scent, and it soothes the skin.

The leaves can also be used for food. They have an astringent quality that pucker up the

Figure 44.7. Prickly rose (*Rosa acicularis*) stems in winter.

Figure 44.8. After and before dyeing with prickly rose.

mouth. I like them dried and crumbled into baking. Young fresh shoots of rose can also be eaten, when they first poke up, before the stems have hardened. When gently steamed and tossed in butter, they make a pleasant side dish.

In midwinter, when I spot the bare stems sticking out of the snow, I like to snip a few red rose branches and cut them into bite-sized chunks to fit my pot. For this I often use prickly rose (*Rosa acicularis*) rather than dog rose since they have a more vibrant red color in the winter. I make a decoction with them, and the water turns dark red. Sometimes I add a few large-leaved avens roots (*Geum macrophyllum*) while decocting for some additional clove-like spicy flavor. I like to drink it warm with honey.

Other uses

There is a reason we see rose used in so many body and beauty products: its astringent nature helps to close the pores and nourish the skin. Rose cleans the skin in a gentle manner and helps dry skin to retain moisture. If I have been out in the sun and my skin feels tight or sunburnt, there is nothing as soothing as some rose water gently dabbed on the skin. The skin becomes relaxed, rejuvenated, and remoisturized. I have only used the petals or the leaves for skin care. The essential oil from the rosehips is said to be beneficial for the skin as well, but I have not tried that. The rose also has antioxidant and anti-inflammatory properties, as well as an abundance of vitamins A and C. The rosehip oil contains retinol, which is recognized as an antiaging agent.

The flower's scent has a calming effect, and I find it to be helpful to use when agitated or anxious. One way to have the scent at hand is to freeze the infused water as ice cubes and store them in a bag in the freezer. When you want to inhale the scent, take out a cube, add it to your water, and enjoy the scent as you drink it.

Rose leaves can be used as a wound healer. If the kids get a small cut while out, we sometimes grab a rose leaf, chew it up into a wad, and then place it over the cut. The astringency in the leaf helps to heal the wound and stops the bleeding.

I have just started to experiment with dyeing small pieces of fabric with various plants. I tried dyeing cotton with rose twigs from prickly rose (*Rosa acicularis*). The process was very easy: I boiled some twigs with the fabric and then

Caution

Do not consume the small hairs in the rosehip. They can cause discomfort to the throat and intestinal system.

Watch out for the formidable prickles, they easily rip the skin. Luckily the petals and the leaves can help with the wounds inflicted by the prickles!

Since this rose is in some places considered an invasive species, pay close attention to where you harvest it. Sometimes people spray them with poison to get rid of them. If you see wilted yellow leaves, or the plant looks off, avoid it.

Figure 44.9. Prickly rose.

Figure 44.10. Baldhip rose.

added a couple of tablespoons of vinegar. I let it sit for 10 hours and then washed the fabric with soap, and the color appears to be sticking. The result was a pale pink purple that I find very attractive. It has an old-timey feel to it. An embroidered rose onto the fabric would complete the picture.

Similar species / look-alikes

There are many wild rose species in the region as well as innumerable variants of cultivated roses for gardens. It is beyond the scope of this book to cover them all. I have not heard of any rose that you could not use in the same way as the dog rose. Some may not be as tasty, but they are still edible.

- **Prickly rose** (*Rosa acicularis*) Deciduous shrub (up to 5 m). Has straight skinny prickles. Both flower and leaf look like dog rose. Hips are more pear shaped and redder. Stems are a deep red in winter. Edible.
- **Woods' rose** (*Rosa woodsii*) Deciduous shrub (up to 3 m). Prickles are slightly bent and sturdy. Stems are often reddish. Flowers and leaf look like dog rose. Hips are more rounded. Edible.
- **Baldhip rose** (*Rosa gymnocarpa*) Deciduous shrub (up to 2 m). Much smaller than dog rose. Hips lack the five dried sepals that most other rosehips have. Hips are tiny and bright red. Leaves are small and dainty. Flower is dark pink. Edible.

Late Fall Oyster Mushroom

Panellus serotinus/Sarcomyxaceae | Mycenaceae

Figure 45.1. Growing on wood.

Synonyms

Late oyster mushroom, olive oysterling, winter oyster, greenbacks, mukitake, green oyster, winter panellus, olive-brown panellus

Description

Saprobic (decomposer) mushroom that lives on recently downed wood. I most often find them on fallen birch with the bark remaining. Appears in late fall (October-December) and can remain until snow. Grow in shelf-like clusters or scattered individually. Mushroom is **fan or kidney shaped.**

Cap is 3 to 10 cm and varies from yellow to brown to green to a purple sheen. Top is glistening and sticky to the touch when fresh. Edge of cap is rolled inward. The orange-yellow **gills** are close together and run down the short stumpy pseudo-stem. It often has small gills that reach partway (secondary gills). The **pseudo-stem** is usually brown/orange, velvety, and often has specks of brown or dull green. Flesh is rubbery and can be waterlogged. **Flesh** is whitish and does not change color when sliced. Does not have a strong odor. **Spore print** is whitish yellow.

Panellus has several meanings, such as "panel, cushion, net or small loaf." *Serotinus* means "late" and refers to the lateness in the season for the fruiting body.

Figure 45.2. Note glistening cap.

Uses

The late fall oyster mushroom is well camouflaged: it looks just like a wet autumn leaf on a log at first glance. These mushrooms tend to be abundant in the cold and dark fall. When these mushrooms start to adorn the woods, I know that culinary mushroom hunting is coming to an end for the year.

I think they are a tasty mushroom if you cook them for a long time. I dry-fry them in a pan to release all their water. Then I add butter and salt and cook them 20 to 30 minutes, sometimes longer on low heat. I like my late fall oyster mushrooms to be crunchy. They are delicious added to an omelet, risotto, or stew. If you do not cook

Caution

Make sure that you pick this mushroom well away from contaminated sites as it absorbs pollutants and toxins such as heavy metals and radioactive waste.

I experienced nausea when I ate overly mature specimens after they had been frozen and thawed in the woods for a few days. I have never had that reaction to a fresh specimen. Make sure to cook it for a long time over low heat as this helps it to be better digested. Do not eat raw.

Figure 45.3. Mushroom omelet.

them long enough, their texture is a bit slimy and rubbery. By cooking, they are broken down so that the human body can more easily digest and access their nutrients. Their flavor is buttery and slightly sour with some bitter undertones. I find them a welcome addition to our fall fare.

Other uses

Some interesting research has been done at Saga University in Japan on this mushroom for its immunomodulating effect. It has shown to have great potential to reduce fatty livers. It also has antitumor substances and has been helpful in combatting malignant tumors/sarcoma.

This mushroom is an effective tool in bioremediation; it absorbs radioactive contamination and other pollutants from areas with exposure. Late fall oyster mushrooms found around the Fukushima Daiichi nuclear power plant in Japan had absorbed and stored huge quantities of radioactive waste after the damage to the plant in 2011.

A close relative of this species, the luminescent panellus (*Panellus stipticus*) was used in the past as a styptic (to staunch bleeding). This species is also bioluminescent; it glows in the dark! Its ability to glow seems to vary over regions, and the mushrooms on the Pacific Coast tend to not have the glowing gene whereas the eastern type does.

Figure 45.4 Grey oyster mushroom.

Figure 45.5. Ruddy panus.

Similar species / look-alikes

- **Grey oyster mushroom** (*Pleurotus ostreatus*) Cap is a lighter color, and the edge tends to be wavy rather than firmly rolled under. It grows on dead trees and fallen logs. Stem is white and smooth (up to 4 cm). This mushroom shows up earlier in the year and is usually not around as late in the season as *Panellus serotinus*. However, if it shows up in the late fall, it can be more rounded and brown, making it look more like *P. serotinus*. Spore print is a pale lilac grey. Edible and tasty.
- **Bear lentinus** (*Lentinellus ursinus*) Brownish dry cap (2.5 to 10 cm). Fan shaped with an in-rolled margin. Grows on dead wood. Prefers conifers in the west and oaks in the east. Lacks a stem. Grow in clusters. Gills are toothed and white pink to brown. Spore print is white. Season is summer to early winter. Not edible.
- **Luminescent panellus** (*Panellus stipticus*) Small brownish cap (1 to 3 cm). Surface of cap has a cracked dried mud look to it. Gills are pinkish brown, narrow, and descend the stalk. Stalk is very short, hairy, and attached off-center. Grows on dead wood in clusters. Flesh is thin and leathery. Spore print is white. Some specimens exhibit luminescence (give off a green light in the dark). Has been used to stop bleeding. Not edible.
- **Ruddy panus** (*Panus rudis*) Covered in velvety hairs. Cap is red/brown to tan with a violet tinge. Stem is short and off-center. Gills are close together and go down the stalk. Grows on logs and deciduous trees in clusters. Spore print is white. Edibility debated.

Large-leaved Avens

Geum macrophyllum | Rosaceae

Figure 46.1. Large-leaved avens seed ball and cauline leaves.

Synonyms

Largeleaf avens, big-leaved avens

Description

Native perennial herb (30 to 100 cm) in moist woods with a mixture of sun and shade. Fibrous **root**. **Stems** are yellow/green and hairy. The stem leaves and the basal leaves differ from one another. Stem leaves (cauline leaves) are alternately attached. **Cauline leaves** are compound with an end leaflet that is divided into three pointy and toothed lobes. Leaflets have serrated edges. Stipules are conspicuous and have large sharp teeth. The **basal leaves** are lyrate, long stalked, and large. End leaflet on the basal leaf is rounded with three lobes and is the largest leaflet. It has several smaller sets of leaflets along the leaf stalk, each set gradually becoming smaller as it approaches the ground.

Flowers (1 to 2 cm across) bloom between June and August. They are yellow with five oval petals and form at the end of branching stems on top of plant. Occasionally petals have a small notch at the tip. Large-leaved avens has five green sepals but most often lack epicalyx bractlets. (See drawing with comparison of *Geum aleppicum* on page 179.) Sepals bend away from the petals. Flower has a green, bulbous center surrounded by anthers with yellow tips. **Seed head** is a ball with hooks radiating out. Each spike/beak of the ball has a hooked tip, and they tend to droop slightly downwards. The surface of each beak has stipitate glands that give it a sticky, glistening look. Each of the hooked beaks contains a seed that travels by attaching itself to passing animal fur or clothing.

The name *Geum macrophyllum* comes partly from *geno*, old Greek that means "to yield an agreeable fragrance." This refers to the odor of the roots when freshly dug up. *Macrophyllum* means "large leaf."

Figure 46.2. Basal leaf.

Figure 46.3. Seeds and seed head.

Uses

Before the ground freezes completely, I like digging up a few large-leaved avens roots. I love using them fresh, but you can dry them too. I clean them well and add them to a pot of water and decoct them. After about 20 to 35 minutes, I strain the roots. I save the water and put the roots aside to use again; I can get two or three batches out of the same roots. The flavor is spicy in a clove-like way. I love drinking it warm with a bit of honey added for sweetness. It is warming and comforting and reminds me a little bit of gingerbread. Avens contain eugenol, which is also found in cloves; it gives the roots their exotic flavor. The decocted liquid would be a good candidate for using to make homemade candy or throat lozenges. See recipe in yarrow chapter on how to make throat lozenges, page 151.

Large-leaved avens roots can be harvested in early spring too; the old wilted leaves from last season will show you where to dig. The reason I prefer to collect them in autumn is so that I can plant the seeds into the dug-up soil at the same time I harvest and ensure that future plants flourish.

The leaves can be eaten too. They are best in the spring. I prefer them cooked to raw because of their hairs. They are not to everyone's liking; they are quite astringent. I like adding them to quiche or empanadas. They can also be dried and crumbled up in bread or soups.

Other uses

I find that the decoction of the roots can settle an upset stomach and soothe a raspy throat. According to the authors of *Ethnobotany of the Okanagan-Colville Indians of British Columbia*

Figure 46.4. Avens flower with prominent sepals and petals with a notch at tip.

Figure 46.5. Fibrous root.

Figure 46.6. Agrimony seedpods.

and Washington, an infusion of the roots was taken by women after childbirth. I have seen other references to this plant's use as help during labor and after childbirth. One report stated that an infusion of the leaves was taken to avoid conception.

The astringent properties of the leaves make it a useful plant to heal wounds or add to bruises. For this purpose, a warm poultice of the boiled leaves can be helpful. A leaf infusion can also help to combat diarrhea.

The root can be tinctured, and it turns the alcohol slightly red. The roots are bruised before placing in alcohol for 14 days.

Mrs. M. Grieve writes in *A Modern Herbal* about a relative, wood avens/clove root (*Geum urbanum*), that can be taken for gastric irritation, fevers, and headache. Wood avens has been introduced into North America from Europe and the Middle East and can hybridize with various other species of *Geum*. In the past, people have used wood avens to flavor ale and

Caution

Do not use while pregnant or while trying to conceive. Some reports speak of its use as a contraceptive. It is also employed to ease labor so could interfere with pregnancy.

Figure 46.7. Agrimony leaf.

preserve linen from attacks from moths. It was harvested on a designated day in early spring, the 25th of March. The reason for that date was that it was said that the root was the most fragrant then.

Similar species / look-alikes

- **Agrimony** (*Agrimonia eupatoria*) Flowers are small and yellow and form on a spike. Seed heads are cone shaped (up to 1 cm). Leaves are similar, but the basal leaves have jagged points rather than rounded lobes. Edible.
- **Wood avens** (*Geum urbanum*) Introduced species. Basal leaves are slightly rounded with three lobes. They are not as large as *G. macrophyllum*. Edible.
- **Water avens** (*Geum rivale*) Basal leaves are less rounded, usually three-lobed. The flower nods, and the purple-brown sepals closely hold onto the flower. Edible.
- **Yellow avens** (*Geum aleppicum*) Very similar-looking species. The sepals differ slightly. *G. aleppicum* has epicalyx bractlets; see drawing. Seed head is more elongated, hairier, and lack the sticky glands. Basal are three-lobed and jagged. Edible.

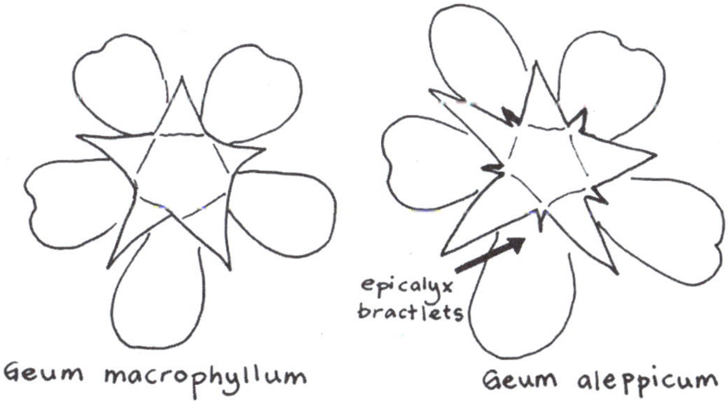

Figure 46.8. Flower comparison: large-leaved avens, *left*; and yellow avens, *right*.

Western Mountain Ash

Sorbus scopulina | Rosaceae

Synonyms

Greene's mountain ash, rowan, dogberry, life-of-man, roundwood, wild ash, quickbeam, winetree, mountain sumac, service tree, Indian mozemize, missey-moosey

Description

Deciduous tree/shrub (1 to 5 m tall). Likes forest edges and clearings with moist ground. It often has several stems. **Trunk** is slender and **bark** is smooth, brown, often with splotches of whitish lichen/algae. **Leaves** are pinnately divided into 7 to 13 pairs of oblong leaflets. Leaflets have toothed edges. Leaves are dark green on top and whitish below. When the leaf and flower **buds** are young, they are covered in white hairs. Buds are sticky in spring. **Flowers** are creamy white (1 cm across). They have five petals and from afar give a fuzzy appearance. Held in large clusters (up to 200 flowers) with flat to rounded tops. **Berry** (pome) is orange red and looks like a small apple. They have small brown, half-moon **seeds** that often hang in the tree throughout the winter and become mushy after frost. Cedar waxwings (*Bombycilla cedrorum*) are attracted to the trees and often arrive in great flocks to eat the berries during winter.

Figure 47.1. Western mountain ash berries in late fall.

The name "mountain ash" is a misnomer. It got this name due to the leaves resembling those of ash (*Fraxinus* spp.). They are not related; ash is in the Oleaceae family (olive), and mountain ash is in Rosaceae, the rose family. *Sorbus* is Latin and means "service tree." *Scopulina* means "of rocky places."

Uses

To be perfectly honest, I find it hard to tell some of the mountain ash species apart from one another at first glance. Not only are there quite a few native species, but various cultivars have also been introduced from elsewhere, and some have escaped into the wild. Luckily, they can all be used in the same manner. Later in the year, the berries are sweeter and juicier. Late fall is when I like to harvest them. At this point, they are mushy and wrinkled and their color has deepened. They are such a joy to harvest! They lighten up the landscape and are easy to pick. The berries form in big clusters so it does not take many to fill a basket.

I use the berries mostly for jelly. In Sweden, rowan jelly is common and used with game meat such as moose or deer. I like making a batch to use with our lamb roasts. It goes well with any meat dish. It has a tart and refreshing flavor that contrasts nicely with fatty meats. While you are making the juice for the jelly, make a double batch to turn some into mountain ash Turkish delight. They are what the name implies: a delight. See recipe page 182.

You can also dry the berries and make flour from them. You can eat a few of the berries raw, but if you consume too many, they will make you sick. Cooking renders them safer. In their raw state, they contain parasorbic acid, which converts into sorbic acid once cooked.

I like harvesting the leaves in the spring. I dry these and use them as flavoring in baking. They have a faint hint of bitter almond/marzipan. They contain some cyanogenic glycosides (which is what gives them that flavor), but cooking or drying destroys its toxicity. The leaves are excellent slightly fermented/oxidized and then redried and used as tea. I dry the leaves flat, away from sunlight in a place with good ventilation for a day. When they are almost completely dry, I separate the leaves and scrunch them up. I sprinkle a fine mist of water on them and

Figure 47.2. Flower clusters in late spring.

Figure 47.3. Bud with white hairs.

Figure 47.4. Western mountain ash leaves in July.

Mountain Ash Jelly

2 cups berries
2½ cups water

- Boil berries, then strain the juice through a cheesecloth overnight. I squeeze the berries to get as much of the flesh as possible. If you want a clear jelly, skip the squeezing.

2 cups mountain ash berry juice
4 cups sugar
½ pouch of powdered pectin

- Cook for 20 minutes. Can or freeze.

My mountain ash jelly is usually a bit runny; if you want it to be thicker, add more sugar or cook for longer. I usually do not make big batches of this jelly; a little goes a long way. It is an acquired taste, and not everybody likes it straight away. My kids like it with butter on crispbread.

Mountain Ash Turkish Delight

1 cup + ½ cup mountain ash juice
2¼ cups sugar
1 tsp. lemon juice
2 tbsp. gelatin powder
½ tsp. vanilla essence
arrowroot flour

- Butter a baking tin.
- Mix sugar, juice (1 cup), and lemon juice in saucepan; bring to boil. Boil until so thick that a droplet of the liquid forms into a hard ball in cold water.
- Remove from stove, let sit for 10 minutes.
- Heat remaining ½ cup mountain ash juice, dissolve gelatin in it, then add vanilla, mix well.
- Pour into mountain ash and sugar mixture, mix well.
- Pour into baking tin and keep in fridge overnight. The following morning, cut into squares and dip in arrowroot flour.

continue to dry them in a dryer with forced air on low heat. Make sure that your batch never has a chance to get moldy. They taste marvelous, like good-quality black tea. Even just dried, without fermenting, the leaves make a decent cup of tea.

Other uses

One time I was so hoarse, I had almost lost my voice entirely, and I was about to give a presentation that required me to talk for a few hours

Figure 47.5. Mountain ash Turkish delight.

Caution

The raw mountain ash berries can cause nausea if too many are ingested. Cooking renders them safe.

The seeds, bark, and leaves all contain cyanogenic glycosides. Drying or cooking removes some of its toxicity.

without pause. I had read that fresh mountain ash juice could help with this ailment. I made myself a large glass of freshly squeezed juice. I gargled with this liquid for several minutes and then swallowed the juice. This is what happened...It helped with my hoarseness, I regained my voice and was able to do the presentation, but...I had the worst cramps as the liquid went through my system. Mountain ash is extremely astringent and as such it worked well to heal my inflamed throat, but it wreaked havoc on my intestines. I could feel every inch where that juice went. So, take it from me, do not ingest that much raw juice. If you want to use it for a sore throat or hoarseness, spit it out after you gargle.

The fruit has been used to help with diarrhea and to combat scurvy in the past. It is high in vitamins A and C. It is also a source of some iron and protein. The berries contain pectin; the more unripe they are, the higher the amount.

Similar species / look-alikes

- **Sitka mountain ash** (*Sorbus sitchensis*) Native tree/shrub (1 to 5 m). Very similar to western mountain ash, it has rust-colored hairs on buds rather than white hairs. Used in the same manner.

Figure 47.6. Common ash.

- **Common ash** (*Fraxinus excelsior*) Large tree (up to 30 m). Has hanging samaras in the fall. The end of the branches has a very distinct black marking, which to me look like deer hooves. The pinnately compound leaves resemble *Sorbus* spp., but ash leaves are typically larger and shinier, and has a larger end leaflet. Can be used for food and medicine.
- **Sumac** (*Rhus* spp.) Small tree (up to 5 m) Large pinnately compound leaves, 9 to 11 pairs of leaflets with serrated edges. Small green/white flowers in panicles that later

Figure 47.7. Sumac.

Figure 47.8. Rowan.

form into reddish hairy drupes. Some parts are edible.

- **Rowan** (*Sorbus aucuparia*) European species. Stem is slender and smooth, often yellow grey when young. This tree can grow tall (up to 15 m). Berries are redder than *S. sitchensis* and *S. scopulina*. Leaves have 4 to 9 pairs of pinnate leaflets. Can be used the same.

Prince's Pine

Chimaphila umbellata | Ericaceae

Synonyms

Pipsissewa, umbellate wintergreen, love-in-winter, wintergreen, noble prince's pine, ground holly, king's cure, rheumatism weed, bitter wintergreen, false wintergreen

Description

Slow-growing **evergreen perennial** with slender woody stems (up to 35 cm). Grows from **rhizomes** in shaded places. **Leaves** are dark green and glossy, oblong or lance shaped, wider at the tip. The leathery serrated leaves grow in whorls of 3 to 8. **Flowers** (usually 3 to 7) bloom in an umbrella shape on a stalk that sticks up above the leaves. Flowers have five petals, white with pink at their bases. The middle of the flower has a prominent green style with ten stamens surrounding it in a halo. In late summer, the **fruit** forms, a dry capsule that divides in five. It is reddish brown and can remain on the plant throughout winter. **Seeds** are tiny beige-brown specks. They are very light, and the slightest breeze stirs them up.

The name *Chimaphila* comes from the Greek *cheima*, meaning "winter," and *phileo*, "to love." *Umbellata* refers to the nodding flowers in umbels.

Important facts!

Because prince's pine is so slow growing, it cannot handle extensive harvesting. **This plant is also incredibly sensitive to trampling.**

Figure 48.1. Prince's pine leaves.

Figure 48.2. Flower.

Figure 48.3. Seedpods.

Figure 48.4. Prince's pine, green and vibrant under the snow.

The plant is partially myco-heterotrophic, meaning that it relies on fungi for part of its nutrient intake and that it produces some food itself by photosynthesis. It creates relationships with both parasitic fungi as well as maintains its own mycorrhizal network. Due to this incredibly complex interconnectedness, prince's pine does not handle ecological disturbances of any kind very well.

This plant likes well-established forests older than 20 to 30 years old and will be absent from recently burnt sites and areas that have been clear-cut. It is an often-seen plant in old-growth forests.

I was initially concerned about including this plant in the book because I worried about the potential risk of over-harvesting and possible damage to the plants if more people came looking for them. I concluded that the greatest risk to this plant is not the responsible foragers who love plants but rather not enough people knowing and caring about this plant. Awareness is the first step toward conservation. Hence, prince's pine is in this book. I trust you fellow foragers to do right by this plant.

Uses

Now in November, prince's pine is one of few green plants around. If there is snow, you can still find them in pockets sheltered from the snow, such as under a fallen tree or next to an old stump. Once the snow is deeper, it is hard to know where to find them. I usually harvest only a couple of handfuls, taking care to remove just a few leaves from each plant. I pour hot water over the leaves (fresh or dried) and let steep, covered, for 10 to 15 minutes. This beverage has a subtle sweet flavor and a delicate scent that I find enjoyable. Extract of prince's pine has been used to flavor candy and soft drinks such as root beer in the past.

The leaves can be dried too, and they keep their flavor well. There are some reports of the root/rhizome being used as well, but since this is more invasive to the plant, I opt to use the leaves. They are also a nice trail nibble. The flavor is refreshing, and it clears your palate. I love chewing on one of these leaves when I am out in the woods.

Other uses

Another name for prince's pine is *pipsissewa*, from the Cree word *pipisisikweu* that means "it-breaks-into-small pieces." This could refer to its use in breaking up kidney stones. The leaves can be used for bladder infections. They are also an effective diuretic. I was interested to learn that this plant is recommended by some as a remedy for long-lingering colds with a respiratory component. It was used in the past to treat tuberculosis. Prince's pine has been used historically in the southern interior of BC for colds and to help sore throats. I use it when I have a raspy heavy chest after the flu. It has also been used as a poultice on aching joints and for arthritis and gout. Since the leaves are antiseptic, antimicrobial, and anti-inflammatory, it can also be used topically for wounds, blisters, and swelling. See caution box.

Similar species / look-alikes

- **Menzies' pipsissewa** (*Chimaphila menziesii*) Similar looking but smaller and slenderer. It only has 1 to 3 flowers. Leaves are less jagged than prince's pine. I am unfamiliar with this plant personally. Since it is quite rare in this area it is better to leave it alone.
- **Striped wintergreen** (*Chimaphila maculata*) Leaves have prominent white veins and are in whorls of three. It has 2 to 5 nodding white flowers. Can be used in same manner as prince's pine.

Figure 48.5. Falsebox.

- **Wintergreen** (*Pyrola* spp.) There are several different species of wintergreen/pyrola. Leaves are not in whorls, and they tend to be more rounded. Flowers are nodding, attached on the stalk in the alternating pattern of a raceme. Some species have medicinal properties.
- **Falsebox** (*Paxistima myrsinites*) Leaves can at first glance be mistaken for one another, and they often grow in the same area. Leaves are smaller and grow opposite on low branches. Flowers are reddish brown and are attached among the leaves on the branches rather than on a stalk towering above. Has medicinal uses.

Caution

Do not use long term; it can reduce mineral absorption in the gut. Large doses can cause nausea, vomiting, dizziness, and diarrhea. The urine will turn slightly green when consumed. This is no cause for concern. Due to its tannin content, it should not be taken with other medications as it can hinder absorption. The bruised leaves on the skin can cause reddening, blistering, and peeling of the skin. Try on a small patch first.

Rocky Mountain Juniper

Juniperus scopulorum | Cupressaceae

Synonyms

Weeping Rocky Mountain juniper, Colorado red cedar, red cedar

Description

Small native evergreen tree with irregular or conical **crown** (5 to 15 m). Likes drier ground and rocky south-facing sites. Grows slowly and can reach a high age (800 to 1,000 years). **Trunk** is gnarly with fibrous bark that shreds as it ages. **Bark** is reddish brown when young and grey at maturity. Often the tree will have several trunks. **Leaves** are blue green and scale-like, closely pressed against stem and slightly overlapping. **Stems** have a square appearance, and the foliage appears lacy.

When young, it looks very much like a common juniper (*Juniperus communis*). Then it has pointy awl-shaped leaves in whorls of three. The tree is most often dioecious (female and male cones on different plants). Yellow-brown **male cones/strobili** form at the tips of the branches. They are papery, dry, and produce pollen that spread by wind. **Female cones/strobili** look like berries and have a white waxy coating; they are bluish purple when ripe. They form at the tips of the branches (September through October). Each female cone contains 1 to 3 round, yellow-brown **seeds. Roots** are deep and wide spreading.

Figure 49.1. Mature Rocky Mountain juniper.

Figure 49.2. Female strobili/cones.

Figure 49.3. Male strobili/cones.

Juniperus is Latin, a combo of *junio* "young" and *parere* "to produce." Together they mean "youth producing" or "evergreen." The name *scopulorum* comes from *scopulus*, an ancient Greek word that means either "of the mountain" or "grows on cliffs."

Uses

I like the very distinct flavor of the juniper berry, but I do not consume great quantities of them because they are so potent. They are aromatic, sweet, mealy, and spicy all at once. I like to suck on a few berries when I am out in the woods. They warm the throat, freshen the breath, and give a jolt of energy. I also pick the berries and bring them home to dry. These berries are ready in my pantry for seasoning stews and roasts. I use them as I do bay leaves or whole peppercorns, a few added for flavor. They give a meat stew a lovely wild flavor; I think they match lamb particularly well. They can also be used steeped as a tea ingredient. Do not use if pregnant. See caution box.

The berries can be used to make a sourdough starter. The natural yeasts for the starter are on the waxy white coating.

On the Swedish island of Gotland, a juniper drink called *Gotlandsdricka* or *drikke* has long historical use. It is made from common juniper boughs (*Juniperus communis*), malt, birch sap, and honey. It is slightly fermented and mildly alcoholic. It has been made since the Viking Age. Juniper (mostly common juniper) is what gives gin its unique flavor. The name "gin" comes from either the French *genièvre* or the Dutch *jenever*, both meaning "juniper."

Other uses

The leaves/boughs can be used in a steam bath to help with a cold. I put several branches into a pot, add steaming water, and lean over it to breathe in the steam with a towel over my head.

Figure 49.4. Young leaves.

The volatile oils rise with the steam and help to clear congestion. I find it to be a great aid in getting over a cold quickly. The leaves can also be made into a pleasant tea with a resinous flavor for chest congestion and cough. It calms me down and makes me slightly drowsy.

The Okanagan-Colville made/make a cleansing wash and a smudge with juniper to sanitize a house after death and to ward off any spirits. The plant is strongly disinfectant. During the plague in the fourteenth century, European physicians held a few of the berries of common juniper (*Juniperus communis*) in their mouths while tending the sick to avoid falling ill themselves. In Tibetan temples, juniper incense is burned to purify the temple and all who enter.

The berries are diuretic and will increase urination. Interestingly, some report the urine to smell violet-like after using it.

The smoke of the leaves repels mosquitoes.

Caution

Do not use juniper if pregnant; it can cause miscarriage. This mainly applies for the essential oil of juniper, but it is best to be cautious even with fresh berries and leaves.

Juniper can alter the blood sugar level; if you are diabetic, be cautious and consult with your physician before using.

The oil can cause rashes on people with sensitive skin.

Do not ingest too much since it can then cause kidney failure, convulsions, and an irritated digestive tract. People with known kidney problems should be cautious with juniper.

The wood is hard and durable. It is sought after in making bows, flutes, and yokes for horses. Because of its rot resistance, it is also a popular tree for making fence posts. It is a very slow growing tree, and some have been recorded to live for over 1,500 years; a few may have exceeded 2,000 years. So, before you decide to cut down one of these trees, think long and hard if it is necessary. In front of you could be one of those ancient wise beings, a part of living history.

The plant is important winter forage for pheasants, mule deer, and whitetail deer. The berries are popular with birds such as waxwings, robins, and other songbirds. They help to distribute the seeds. The seed coat is dissolved while passing through the birds' digestive tract, and this increases success with germination.

Figure 49.5. Common juniper in September.

Figure 49.6. Rocky Mountain juniper left, western redcedar right.

Similar species / look-alikes

- **Common juniper** (*Juniperus communis*) Shrub in exposed areas. Grows low to the ground or as tall as 16 m. Leaves/needles in whorls of three. Each needle has a white stomatal band. Berries/strobili take two years to mature and go from green to light blue to dark blue/black. They have a whitish waxy coating. They are more rounded than the berry/cone of *J. scopulorum*. Edible and medicinal.
- **Western redcedar** (*Thuja plicata*) Very large tree (up to 70 m). Wide trunk (up to 4 m in diameter). Bark is reddish and stringy. Leaves are scalelike and highly aromatic when crushed. They are bright green with whitish stomatal bands below and grow flat on the branch. Cones are small and woody. Tree has many varied uses. See western redcedar chapter on page 216.
- **Yellow cedar** (*Cupressus nootkatensis*) Medium-sized tree (up to 24 m tall). Grows in the coastal regions of the Pacific Northwest. Bark is thin and grey-brown, scaly with age. Leaves are scaly, bluish green, and cover the twigs on all four sides rather than being flat, like the red cedar. Berries/cones look like juniper berries when young but are woody at maturity with umbrellalike scales. Wood is lightweight and durable. Has been used for an impressive range of material among BC coastal Peoples.
- **Eastern white cedar/arborvitae** (*Thuja occidentalis*) Evergreen tree native to eastern Canada and northeastern US. Red-brown bark. Leaves are scaly, yellow green to green, and flattened on the branch, giving it a lacy look. Cones are slender with 6 to 8 overlapping scales. Has medicinal properties.

Field Pennycress

Thlaspi arvense | Brassicaceae

Synonyms

Bastard cress, stinkweed, fanweed, frenchweed, mithridate mustard, pennycress

Description

Winter annual (up to 60 cm tall). Grows in weedy places, croplands, and roadsides. Starts as a **basal rosette** with arrow-shaped, narrow, and toothed leaves. As the plant matures, the basal rosette wilts and disappears. Lower stem **leaves** are sessile or have a short petiole. Leaves further up the stem are smaller, alternately attached, and have a pair of lobes that clasp the stem. **Stem** is ribbed and fibrous. **Flowers** form in erect racemes with a rounded cluster of flowers at the very tip. Each tiny flower has four white petals and four green sepals. They have six yellow-green stamens surrounding a flat ovary. Flowers bloom May-July and can self-pollinate.

Seedpod (silicle) is round and has two cells with up to eight seeds in each. The tip has a notch. **Seeds** are oval, brown/black with concentric circles that resemble a fingerprint. Pennycress usually germinates in the fall and overwinters as a small rosette. Has a **taproot**.

The name "pennycress" comes from the fact that the seedpod resembles an old penny. *Thlaspi* is from the Greek *thlaein*, meaning "to crush." It refers to the flattened silicle. *Arvense* means "of farmed and cultivated lands."

Uses

While field pennycress is not my favorite for most of the year, in December I praise it. There are very few greens available to pick at this time of year, so I feel joyful when I see the persistence of the pennycress. I like picking a little of it to eat raw, and I add some to cooked dishes. It has a very strong flavor, like most of its mustard relatives, so only a little is needed.

Glucosinolates are part of all the plants in the Brassicaceae/mustard family, and they are responsible for those bitter and pungent flavors. I do not recommend eating too much of it since it can cause stomachache. See caution box.

Figure 50.1. Pennycress seedpod (silicle).

Figure 50.2. Flower.

Figure 50.3. Leaves on stem. Note clasping lobes.

Figure 50.4. Seeds in opened silicle.

Figure 50.5. Basal rosette in spring.

I also like using the seeds. Most of them have been scattered in December, but a few remain on the plant. I gather these and use them as a spice in cooking. They taste like mustard seeds. I find that they give me energy and jolt me awake. I like sprinkling a few on a pickle sandwich. The ripe seeds can be ground into a paste and turned into a prepared mustard with vinegar, sugar, and salt. When I have attempted this, the result has been too potent for most palates. The seeds are lovely sprouted and added to sandwiches or salads.

Earlier in the year, while the seedpods and seeds are still green, they can be added to a salad for a spicy crunch. They look cute, and their vivid green looks very attractive in a potato salad. Plant is high in vitamins C, B2, and sulphur.

Caution

Contains glucosinolates that, if ingested in too high doses, could result in a decrease of white blood cells, nausea, and dizziness. Do not consume if pregnant or breastfeeding.

There are cases when cattle ingested large quantities of pennycress in hay (25 to 100 percent) and become sick, aborted their young, or even died. During dry spells, this seems to be more of a concern. There are some reports that hay turned into silage converts the allyl thiocyanate in the pennycress from being poisonous to being safe for the cattle.

Other uses

Chewing the leaves into a wad and placing on a bug bite/sting can help with pain or itching. Traditionally pennycress has been used as an anti-inflammatory, a blood tonic, and blood purifier. I have not tried it for any of these purposes myself. It can be used to remove phlegm/mucus, and as such, I find it to be quite effective. When I feel stuffed up, I sometimes take a few seeds and either chew them raw or make a tea with them. This loosens phlegm and makes it easier to expel. The tea also helps with a sore throat. It has antimicrobial activity against staphylococcus and streptococcus. The entire plant can be used to reduce fever. Seeds and young shoots are supposedly good for the eyes. The plant has also been used as a remedy for arthritis.

Another interesting use for pennycress is as a component in conventional agricultural systems. It protects the soils over winter and early spring by preventing erosion and nutrient loss and keeping other weeds at bay. The plant is under research as it produces an oil that can be used as biofuel. Perhaps one day you will be driving your car with zero emission thanks to the weeds in your garden. Now, that is a wonderful thought!

Similar species / look-alikes

- **Shepherd's purse** (*Capsella bursa-pastoris*) Has triangular heart-shaped seedpods with the narrow end at the base (obcordate). Leaves are more pinnately lobed than pennycress. Often grows next to one another. Edible.
- **Wormseed mustard** (*Erysimum cheiranthoides*) Has elongated seedpods (siliques) rather than rounded ones. Flowers are yellow. Leaves are very similar and can be

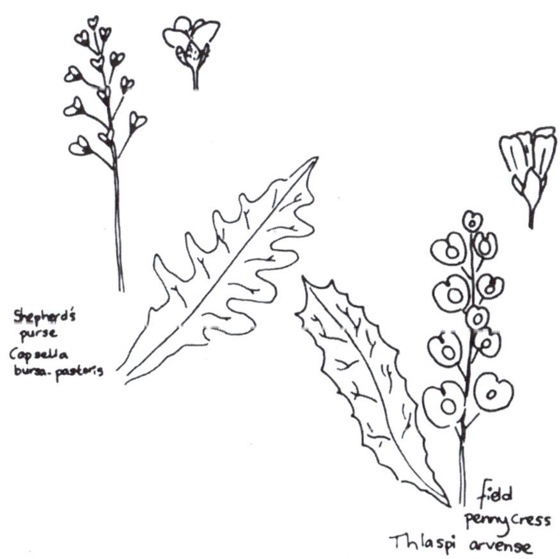

Figure 50.6. Comparison shepherd's purse, *left*; field pennycress *right*.

mistaken for one another in their basal rosette stage. Has an acrid metallic taste that I find unpleasant.

- **Hoary alyssum** (*Berteroa incana*) White flowers with four petals. Each petal has two lobes. Leaves are blue green, entire, and hairy. Seedpods are oval. Not edible. **Very toxic to horses.**
- **Hoary cress** (*Cardaria draba*) Leaves are very similar in both color and shape. White flowers are borne in larger flat-topped clusters. Seedpods are heart shaped and smaller than pennycress. It has a creeping underground root system. Edible with caution.

Trembling Aspen

Populus tremuloides | Salicaceae

Synonyms

Quaking aspen, American aspen, golden aspen, trembling poplar, mountain aspen, white poplar, popple, aspen poplar, smooth-barked poplar

Description

Native deciduous tree (up to 40 m tall) that most often grows in stands with many other trembling aspens of similar age. Fast-growing pioneer species that comes back quickly to populate recently burnt areas. Cannot handle dense shade or prolonged drought. Individual trees live on average 50 to 75 years but can live to be 150 years old. **Trunks** are slender and blue grey with black markings (horizontal black rings and black knots) when young, they become thicker barked and furrowed at base with age. **Bark** does not peel and is covered in white powder. **Root system** is elaborate, and the tree can clone itself and grow new saplings from its underground lateral root system.

Branches are grey brown and have pointy winter **buds** that get sticky with resin/oil in early spring. Buds have three or more scales that overlap like shingles. Wood and buds have a distinct perfume/resinous smell. **Leaves** are alternate and widely ovate when mature, 4 to 8 cm long (width slightly shorter). Young sapling leaves are larger, around 10 to 12 cm, and less rounded. Each leaf is attached to a long petiole that is flattened; this allows the leaf to move in the slightest

Figure 51.1. Trembling aspen in winter.

Figure 51.2. Buds in February.

Figure 51.3. Male flowers on catkins.

Figure 51.4. Young sapling. Note that the leaves are larger and a slightly different shape than mature leaves.

Figure 51.5. Mature leaf: close-up of petiole (leaf stalk). Note the flattened shape.

breeze, hence the name "quaking" or "trembling aspen." Leaf margin is crenate-dentate (somewhere between scalloped and toothed). Top of leaf is matte blue green; bottom is pale green to white. Leaves turn yellow in the fall.

Flowers in mid-spring. Male and female flowers are on different trees (dioecious); they are pink. Male flowers are on shorter catkins, and female catkins later produce tiny **seeds** encompassed in cottony fluff. Wind-pollinated, and dispersal of its seeds is by wind or water. The seeds do not remain viable for very long after leaving the mother tree.

Populus means "people," and *tremuloides* means "resembling the quivering poplar." This refers to the closely related European aspen (*Populus tremula*).

Uses

The kids like to nibble on the buds during winter. They taste refreshing and interesting. While I would not eat too many of them, they are a nice treat occasionally.

Aspen Bark Sourdough Starter

- I add a handful of bark pieces to 2 cups of flour and 2 cups of lukewarm water. I let this sit by the wood stove for a day or two (any warm place will do). After a day, it starts to bubble away. A large tablespoon of the starter for a large loaf in place of regular yeast seems to do it. I let the dough sit for an extra day before baking it. I have not tried using this sourdough starter in a gluten-free loaf, so I do not know if it would work as well. With regular flour, it rose nicely and made a decent loaf. If you use too much of the sourdough starter, the bread will have a very strong flavor. The first time I tried it, I made it so potent tasting that no one else in the family wanted to eat it!
- Keep feeding your sourdough with fresh water and more flour as you take some out to use. After a while, mine developed a hard crust on the surface, and I said good-bye to it. I have a feeling that I might have neglected it on a few too many occasions. You know how it is...

Often little aspens are toppled by heavy snowfall. I like shaving off some of the bark of these trees. You do not need a lot to start an excellent sourdough, a handful is enough. The white powder on the outside is where the wild yeasts are located. This yeast will rise your bread.

Figure 51.6. Bark for sourdough starter.

Figure 51.7. Aspen bark sourdough bread.

Other uses

The buds can also be collected and added to oil; I use olive oil. I gently heat up the buds in a double boiler with just enough oil to cover them. A handful of buds and 1 cup of oil made enough to last me for a year. The oil should just be heated, not allowed to boil. I let the buds cool in the oil and leave them in there for a month. I use this oil on sore and tired muscles or as a chest rub if congested. It feels good on chapped hands, feet, and lips.

The tree bark contains salicylic acid and can help with fevers and pain. Be cautious when giving this to a child. See caution box.

The powder on the bark can be rubbed on the skin as sunscreen. It works ok. I tried it on half of my face, and that side was slightly less burnt than the other side (with no sunscreen) after a day in the sun.

Aspen is an important native tree that feeds many species. Numerous animals rely on its twigs in the winter for food and shelter. Elk, deer, vole, beaver, ruffed grouse, squirrels, and rabbits are a few of the critters that appreciate the aspen groves. Leaf borers such as the aspen leaf beetle (*Chrysomela crotchi*) and the cottonwood leaf beetle (*C. scripta*) love the leaves. Many other beetles bore through the wood, and they attract various birds and woodpeckers. Outside my kitchen window, I have a few dead aspens, and they are visited by pileated woodpeckers, downy woodpeckers, red-breasted nuthatches, and northern flickers. The beautiful Compton tortoiseshell caterpillar (*Nymphalis vaualbum j-album*) feed on the aspen leaves, as do many other butterfly and moth caterpillar species. These trees keep entire ecosystems alive!

Aspens are interesting in that they can clone themselves. They produce seeds but can shoot up a new tree from an older root, making a genetically identical clone. In Utah, an aspen grove known as Pando has been estimated to cover 100 acres and weigh as much as 40 blue whales. Every tree in the grove comes from one tree; all these clones of the mother tree are considered to be one organism. Now that is one big being! In fact, it is considered the largest living thing on Earth. It is estimated to be 80,000 years old, making it the oldest organism as well. The sheer enormity of these facts blows my mind away! Now this grove is under threat. It has suffered from rising temperatures and droughts over the last years, and it is not regenerating as it should to survive. Mule deer graze on it, and because the top predators are gone, they are not kept in check. As a result, all the young saplings get eaten before they can become mature trees, and the grove is destined to die out if no new growth can succeed. Every creature in the forest has a role to play, and we humans have disrupted that. With our sprawl across the land, we have caused so many species to become at risk of extinction due to habitat loss. When there is no room for apex predators such as wolves, bears, and cougars, every species down the food chain suffers eventually, including humans. Balance needs to be restored. When you think to landscape your surroundings, keep the native species and have a gentle touch with the land, leave habitat corridors for animals to traverse through. Everyone will benefit.

Caution

If you have an allergy to aspirin, do not use aspen bark. If you are pregnant or nursing, check in with your doctor before using. Do not give aspen bark to children under sixteen when they have the flu, chickenpox, or a viral infection. They can develop Reye's syndrome which is a very serious, albeit rare, disease.

Figure 51.8. White powder on aspen bark.

Similar species / look-alikes

- **Paper birch** (*Betula papyrifera*) Bark looks similar but will flake and peel; aspen bark does not. Leaves are more triangular, and petiole is not flattened, but rounded. Fruit is a winged nutlet; catkins do not have cottony fluff. Edible. See paper birch chapter on page 242.
- **Black cottonwood** (*Populus balsamifera* ssp. *trichocarpa*) (up to 50 m) Grows near rivers and floodplains. Bark is deeply furrowed and grey on older specimens, yellow grey and smooth on younger trees. Leaf is large and dark green and shiny on top. Bottom of leaf is pale green to almost white. The margin is gently dented. Has large catkins with an abundance of cottony fluff. Has many uses. See black cottonwood chapter on page 260.
- **Balsam poplar** (*Populus balsamifera* ssp. *balsamifera*) Leaf is more elongated than black poplar. Balsam poplar and black cottonwood sometimes interbreed and create new hybrids. Can be used in the same way as black cottonwood.

Figure 51.9. Compare the flaky bark of birch with the smooth bark of aspen.

Figure 51.10. Leaf comparison: birch, *left*; aspen, *right*.

Ponderosa Pine

Pinus ponderosa | Pinaceae

Figure 52.1. Distinct ponderosa pine bark.

Synonyms

Bull pine, blackjack pine, western yellow pine, filipinus pine, western red, western pitch, Sierra brownbark, western longleaf pine, pino real, western yellow, black jack, Washoe pine

Description

Stately evergreen tree (up to 80 m). Grows in dry areas between grasslands and forest. Can handle drought; mature species can survive moderate fire, thanks to self-pruning, high branches, and thick bark. Thick **bark** is very distinct: it forms in plates like a jigsaw puzzle. It is bright orange/yellow and red/brown with dark black markings between the plates. Bark is strongly scented on a warm summer's day; I think it smells like vanilla. **Branches** are high in the tree and carry the needles. **Needles** are in bunches (fascicles) of three. Each blue-green needle is around 7 to 25 cm long. Where the needles come together in the bundle, they are flattened. The surface of the needle has a lighter line of stomata (pores for gas exchange). Needles are held together with a persistent fascicle sheath and remain on the tree for 2 to 7 years before dropping off.

Tree is monoecious, with both male and female cones. **Male pollen cones** are yellow or red (1.5 to 3 cm long). **Seed cones/female cones** take about 2 years to reach maturity. Mature female cones are 5 to 15 cm long. They start off as small pink-purplish cones and grow to develop spiky brown scales that hold the seeds. Each scaled tip has a prickly umbo (protrusion) that points outwards. Under each scale sit two light-brown winged **seeds**. Each pine cone can bear up to 70 seeds. Ponderosa pine can produce seeds as early as 7 years old and can continue seed production for 350 years. Tree tends to have a normal seed crop every 2 to 3 years with a more abundant seed production every 8 years on average. **Roots** are extensive: they can extend laterally for 45 m and down 2 m. In dense stands, the roots are generally as big as the crown.

Since the tree can handle less nitrogen and phosphorus than other trees its size, it can grow in soils that are not fertile enough for other species.

Pinus is Latin for "pine," and *ponderosa* means "heavy" or "ponderous."

Figure 52.2. Needles in bundles (fascicles) of three, attached to the tree with a fascicle sheath.

Figure 52.3. Immature female cone.

Figure 52.4. Male pollen cone.

Uses

I think that ponderosa pine is the epitome of southern parts of the Okanagan Valley. Ponderosa pine's interesting bark and graceful needles on grassy hills, overlooking green-blue lakes, is what I see in my mind's eye when I imagine an Okanagan summer. The scent alone reminds me that I am no longer in Sweden but in this exotic place where the trees smell like vanilla and the sunflowers below smell like chocolate. Delightful!

In fact, vanillin flavoring can be extracted when the tree is made into pulpwood. We have ponderosa pine on the farm, and our location is a bit unusual for it since the ponderosa pines rub shoulders with western redcedars here. They usually do not like the same habitat.

We harvest the needles year-round for their delicious tea. It is refreshingly lemony with a

Figure 52.5. Seeds.

Figure 52.6. Barb on cone.

hint of resin and makes your breath pleasant. It is full of vitamin C. Do not consume if pregnant. See caution box.

The seeds can be eaten as you would commercial pine nuts. They are smaller but taste very similar. They are covered in a papery sheath that is best removed before popping in your mouth. I harvest them in two ways: one method is to place a blanket or tarp under a large tree and wait for the seeds and cones to fall on it and try to collect them before the squirrels do. The other method works if there has been a big windstorm and cones have fallen to the ground. Take the cones inside and let them open and release their seeds in the cozy heat of your house. They are high in protein and taste delicious.

The inner bark/cambium of the tree has been used for food by Native Americans and First Nations throughout its distribution. It is usually harvested in late winter or early spring but can be collected at any time. When I take inner bark/cambium off a tree, I cut a skinny strip on one side only of the trunk. You never want to ringbark (make a cut that goes around the entire trunk) a tree as this will kill it. With time, the vertical strip will heal, and the tree will recover; however, while the sapwood is exposed, it could become a target for fungus, insects, or other pathogens that could in the long run be the demise of the tree. I do not harvest inner bark from a tree that is already stressed by drought and insects. Always keep the best interest of the tree in mind.

The cambium is chewy when eaten raw, and it takes a long time to make it soft enough to swallow. It reminds me of a very tough piece of meat that you can chew and chew and chew.

The flavor is a little chalky but nice. I enjoy the flavor of it and feel that the cambium is rather filling. I would caution you to drink lots of water with it as it can make you constipated if you consume too much.

I like using the outer bark of the ponderosa pine to make crackers. This recipe is inspired by Sami Elder Laila Spik from northern Sweden. Her recipe uses Scots pine (*Pinus sylvestris*). I modified the recipe to be used with ponderosa pine. The good thing about using the outer bark as opposed to the inner bark is that it is relatively harmless to the tree. Ponderosa pine has thick, almost puzzle-like outer bark. I break off some of these pieces from a few trees until I have a container full. Make sure you do not harvest too much from any one tree. I break the pieces up by hand, as small as I can, and roast them in the oven for 35 minutes at 350°F (180°C). It smells wonderful. Once that is done, I grind them to a fine flour with a coffee grinder (a grain mill would work even better, I think) then sift it to remove any hard bits. With those preparations done, it is time to combine the ingredients and get baking! See recipe on page 206.

Figure 52.7. Outer bark "puzzle pieces."

Other uses

Ponderosa pine has many medicinal uses and is used by both the Syilx and the Secwépemc. The sap/resin can be chewed as a gum and applied to sores. It can also be added to medicinal salves for rheumatism and aches. I like using the resin/sap when I have a sore throat, mixing it with tea made from the needles. It reduces infection, soothes a sore throat, and provides a healthy amount of vitamin C to boost the immune system. I love that this tree is giving abundantly of its medicine in the middle of winter when it is hard to harvest other green plants.

Ponderosa pine contains resins that can be used to make turpentine, rosin, and pitch. The pitch can be used as a fire starter, to waterproof materials, and as a wood adhesive.

Over 200 species of insects like to eat ponderosa pine. It is also an important habitat for mice, chipmunks, and tree squirrels. Ponderosa pine forests are prime habitat for deer and usually have an open understory with grasses and berry bushes such as saskatoon.

The trees are well equipped to handle moderate fires. They have thick bark and can survive some scorching; their crowns are high, out of reach from ground fires. They can self-prune their branches to keep safe. The cones are adapted to open and drop their seeds after a fire. The seeds that go through a low- to moderate-intensity fire grow more quickly as saplings and gain nutrients from the ashes. This cycle of regeneration only works if the fires are moderate.

Ponderosa Pinebark Crackers/ Knäckebröd

¾ cup pine bark flour
3 cups whole wheat flour
1 tsp salt
1½ cups cold water

- Preheat oven to 350°F (180°C). Mix all ingredients together.
- Roll out very thin and prick with a fork.
- Cut into cracker squares and place on greased oven tray.
- Bake 3 minutes (swap your trays half-way through: top to bottom, bottom to top).
- They will harden as they dry. The crackers have a unique flavor. I adore it, but some people find it too aromatic. They taste great with cream cheese or hard cheese. Drink lots of water with the crackers as they can make you constipated.

Caution

Do not consume needles if pregnant; they can harm the unborn fetus. Isocupressic acid in the needles have caused abortions in livestock after they consumed a critical amount of the fallen needles.

Can cause dermatitis in sensitive individuals (resin, wood, and ash). The needles contain terpenes that can have a detrimental effect (after the rain has washed the terpenes off the needles and into the soil) on certain vegetable crops such as wheat.

Figure 52.8. Ponderosa pine crackers, *knäckebröd*.

Larger, hotter than normal fires do damage to the trees and their seeds. Under Traditional Indigenous Forest Management, the understory was burned at regular intervals. Since fire suppression policies were implemented, a lot of fuel piled up on the ground and many species have crowded the ponderosa pine stands. When a fire comes through an area with built-up fuel on the ground, it becomes hotter and uncontrollable faster and can cause great damage to both communities and forests. Ponderosa pine can handle drought and hot summers, making them perfect in the near desert conditions of some parts of the Okanagan. As summers increasingly become warmer, this might be a species that will find a foothold in areas that have historically had more precipitation. It can also handle freezing temperatures down to −40°F (−40°C).

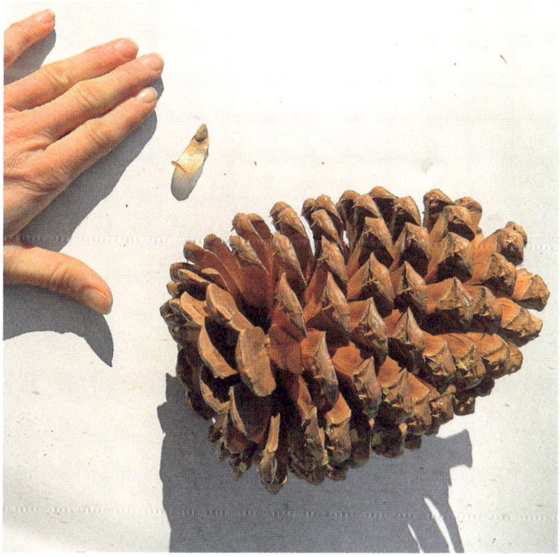

Figure 52.9. Jeffrey pine cone.

Figure 52.10. Ponderosa pine cone, *left*; lodgepole pine cone, *right*.

Similar species / look-alikes

- **Jeffrey pine** (*Pinus jeffreyi*) Grows further south, common in California (up to 50 m). Needles are also in bundles of three. One way to tell Jeffrey pine and ponderosa pine apart is to look at their cones. They both have barbs at the tip of the cone scales. Jeffrey pine's barbs point inwards, making it non-poky to hold in your hand, whereas ponderosa pine's barbs point outwards, making it poky to the touch. To remember, some people use this memory device: "gentle Jeffrey and prickly ponderosa." Jeffrey pine has smaller bark plates than ponderosa pine. Can be used in similar ways.
- **Lodgepole pine** (*Pinus contorta*) Much smaller tree. Bark is grey and does not have large bark plates like ponderosa pine. Needles are held in bundles of two. Edible. See lodgepole pine chapter on page 221.
- **Limber pine/Rocky Mountain white pine** (*Pinus flexilis*) Needles are in bundles of five. Cone is large and cylindrical (7 to 20 cm long).

Figure 52.11. Ponderosa pine needles, *left*; western white pine needles, *right*.

- **Western white pine** (*Pinus monticola*) Long needles (up to 12 cm) in bundles of five. Slender long-stalked cone (10 to 25 cm long) with brittle scales.

Figure 52.12. Laura Dohla.

Figure 52.13. Pine needle basket by Laura Dohla.

- **Whitebark pine** (*Pinus albicaulis*) Lives at high elevations. Needles in bundles of five. Cones lack prickles. Is decreasing in population and is considered endangered. Do not harvest.

Making pine needle baskets with Laura Dohla

Laura Dohla is the Indigenous Education liaison at Highland Park Elementary in Armstrong, BC, and she has worked within the school district for twenty years. Her role in the school system is to share the culture and history of Indigenous Peoples. She works as a liaison between the school, the families, and the local Indigenous communities. Laura is highly passionate about what she does, and when she interacts with the children, you can see what great impact she has. Laura says that the knowledge is not just from her but from the entire Indigenous community. She has had many mentors showing her the way. She has worked with numerous Elders from different Indigenous communities. Juliana Alexander, a Splatsín Elder (Secwépemc), has mentored Laura on how to make traditional pine needle baskets. Ponderosa pine is called *s7etqwllp* in Secwépemctsín.

Laura worries that the language and the knowledge about how to live on the land will be lost if we do not gather as much information as we can from the Elders. The residential schools severely damaged how the culture was passed on to the next generation. (See sidebar on page 209 for more information.) This break in teachings has resulted in the loss of both language and invaluable skills.

Laura explains how she soaks the pine needles in warm water for a couple of hours, and then she pulls off the little black bud sheath from the needle bundle. Other people leave them on and arrange them to meet in a spiraling pattern around the basket. Traditionally, sinew was used in the baskets, but a waxed thread works too. Laura starts by wrapping a bundle of three needles in thread. She covers it all up initially to create strength for the middle base of the basket. To attach another bundle, you simply lay it end to end with some overlapping where they meet. You continue to wrap. The tricky part is bending it to create the circle. That is why the needles need to be soaked; if not, they will be brittle and might snap. Once you have your bend, you use a sewing needle to attach it to the layer before. You go through the middle and back to the row you are working on, pulling tight to secure it. You want to tighten it as much as you can without breaking it. When the needles dry, they will shrink a little, and if your stitch is not tight, the basket will become loose and less sturdy. You continue in this fashion with each coil, attaching it to the previous row with a stitch. Once you have created the middle of the base of the basket, you no longer cover the entire needle bundle but rather make sure your stitches are uniformly spaced for the rest of the basket. The baskets were traditionally used to store berries, dried fruit, or other everyday items. Often, they had lids too.

Laura uses the fresh needles, but others collect fallen needles for their baskets. The needles can be picked any time of year. The longer they are, the easier they are to work with. To keep the basket for longer periods, it can be coated lightly with beeswax, but it is not always necessary. The lingering scent of the pine needles from the finished basket is lovely.

Residential schools: school system in Canada, officially operated from the 1880s to the end of the 20th century. The system forcibly removed children from their families for long periods of time and did not allow them to acknowledge their Indigenous heritage and culture or speak their own languages. The schools were government sponsored, run by religious organizations, and had as their mission to assimilate Indigenous children into Euro-Canadian culture. Terrible atrocities were committed in them, and many children succumbed. The residential schools have left much pain and intergenerational trauma within First Nations communities across the country.

In the US, Native American Boarding Schools or Indian Boarding Schools were in effect between 1860 and 1978. They were a part of the US Native American Assimilation era that began in 1819 when the Congress passed the Civilization Fund Act. Its purpose was to assimilate Native Americans into a Euro-American culture and take their territories. They did this by systematic destruction of Native language, culture, and family. Children in Native American Boarding schools suffered abuse and neglect, and many perished.

Traditionally, women would gather the pine needles for the baskets. Before the women picked the needles, they had to make sure they were in a good place mentally and spiritually. They said prayers to the creator (Tqeltkukwpi7) and smudged. The smudge could be made from cedar or sage. Sometimes it was sweetgrass, which was traded for. Before picking the needles, an offering to the tree is made, often this can be cedar, tobacco or a kinnikinnick smoking mixture. Each step of the process was traditionally accompanied with a specific song. The songs were to honor the tree and ensure success in the making of the baskets. Many of these songs have been forgotten; many songs are being lost as the Elders who still know them pass away.

Laura reminds us that being on the land and learning about the plants and the animals is a crucial aspect of Indigenous knowledge. Being outside, surrounded by nature, and learning by being immersed in the material they have gathered from the wild is a lesson that the children will remember for a long time. It is a different path than the academic route with knowledge gleaned from books. Both ways of learning are important, but Laura feels like there is a need for more hands-on land-based learning in today's school. She has been working with a team to create an outdoor classroom so children and teachers have a space where they can interact and learn from the land. Together with Laura, the students get to learn how to make willow dolls, how to recognize Indian potato (spring beauty) and avalanche lily, and how to use poplar buds, among other things.

Often Laura will invite Elders to come and speak to the children. Their role as mentors and role models, and the knowledge that they have, is paramount in keeping traditions alive. Elders play an important part in Indigenous learning models. In order to teach a culture effectively, you need someone who has a greater understanding of what that culture means, and someone who can explain/show the next generation, not just the historical facts but the way to live the culture. The stories and the teachings are invaluable and teach people not only about themselves but also about how to interact with one another and the land respectfully. The Secwépemc and their ancestors have been here for at least 10,000 years. Much wisdom has been accumulated, and it is crucial that this wisdom is not lost.

If you are curious about learning about plants and culture from Laura, an Ancestral Heartbeat Tour might interest you. Ancestral Heartbeat Tours is a local Indigenous-owned and -operated business in Enderby. They hold tours in Enderby Cliffs Provincial Park (Tplaqin). For more information, contact: AncestralHeartbeatTours@gmail.com.

White Spruce

Picea glauca | Pinaceae

Synonyms

Single spruce, skunk spruce, northern spruce, cat spruce, yellow spruce, western white spruce, Alberta spruce, black hills spruce, Canadian spruce

Description

Evergreen conifer tree (up to 48 m). Can live up to 400 years. Grows in moist soil at lower elevations. **Shallow root system** formed of small rootlets close to the surface of the ground. Topples easily in high winds. **Trunk** can have a diameter of 1.2 m. **Bark** is grey and covered in thin scales. **Twigs** are slender with scales and resinous buds. **Leaves/needles** (1.2 to 2.5 cm long) are stiff and blue green with sharp-pointed tips and stomatal bands. Needles remain on tree for up to ten years. **Flowers** in May-June; male and female flowers on the same tree. Male flowers are pale red, turning yellow with age. Female flowers have scales and are red to yellow green. Female flowers form into cones in fall. **Cones** are cylindrical, pendant, and brown when mature. They have flexible thin scales with a rounded or straight margin. Inside each scale sit two **seeds** with a papery sheath that assists in transporting the seed on the wind. Scales open at maturity to release their seeds. In wet weather, the scales remain closed. The white to pale-yellow **wood** is soft and straight grained.

Figure 53.1. Mature white spruce tree.

Figure 53.2. Needles in winter. Note the sharp tip and light stomatal bands.

Figure 53.3. Fresh spruce tips in May.

Picea comes from the Latin *pix* or *picis* and means "pitch-pine." *Glauca* is Greek and means "blue green" or "blue grey."

Uses

In December, the spruce needles are a welcome source of vitamin C. The tree keeps its needles throughout winter, and they make a lovely tea, with a lemony/resinous flavor and smell. It freshens my breath and gives me energy. The needles in early spring are a real treat. Each spring, fresh needles will form at the tips of the branches. They are soft and refreshingly sour for a short while. During this time, we harvest them to eat fresh, or use to make marmalade or gummies.

The seeds can be eaten, but they are very small. Some years, I find them scattered on top of the hard crusty snow in great quantities, and they can then be picked with ease. I grind them up, husk and all, and add to flour for baking. Make sure not to harvest too many of them as they are an important food for the squirrels and birds that stay for the winter.

The inner bark/cambium has been used historically as food. Care must be taken to not ringbark the tree when harvesting the cambium. The tree can survive having a small section of inner bark taken but may become weakened from this practice and succumb to disease or bug attack once exposed.

My kids often harvest the hardened resin and make it into chewing gum. The trick to a successful spruce chewing gum is to choose resin that is not too sticky. If you harvest it while it is still sticky, it will coat your teeth and flavor everything you eat for the rest of the day.

Other uses

The fine skinny roots of this tree are great for lashing things together and sewing up birch bark baskets. Their combined rigidity as well

Spruce Tip Gummies

1 cup lemon juice
½ cup water
¼ cup gelatin
2 to 3 tbsp honey or maple syrup
½ cup spruce tips, finely chopped

- Put juice, water, gelatin and honey in a pot and heat gently until gelatin dissolves.
- When the mixture is hot (make sure that it does not boil), remove from heat and stir in finely chopped spruce tips.
- Pour into a glass container and put in the refrigerator.
- When it has cooled and solidified, cut into desired shapes.

as suppleness make them perfect for these purposes. To locate the roots, gently rummage in the topsoil by the tree trunk. You will find skinny but strong roots radiating out. Do not harvest too many from one single tree and be careful as you pull them up, making sure that other roots in the area are not snapped off. I scrape off the bark and soak for at least an hour before using.

The resin can be used as a component in salves for cracked skin. Caution is advised. Many people are sensitive to this resin on their skin and can become itchy. My sister was impressed how quickly the salve healed the cracks on her hands, but after prolonged use, realized she was allergic to spruce and had to use an antihistamine cream to stop the allergic reaction. My advice is to try on a small patch first, and if

Figure 53.4. Spruce tip gummies.

Figure 53.5. Spruce seeds and mature cone.

Figure 53.6. Male flowers/pollen cones.

Figure 53.7. Western hemlock.

you feel any redness or itchiness, stop using it immediately. I also harvest the resin and store it in my pantry for when someone has a cough or a sore throat. I throw a chunk into hot water, let steep, and drink this beverage warm with honey. It is great to add the needles to the tea too. They have antiseptic properties.

The wood of spruce is lightweight and soft to work with. It is excellent for making outdoor equipment such as canoe paddles and skis.

Caution

Some people react adversely to the resin. It can cause blisters, rash, and itching. Try a small amount the first time. Take care to not mistake this species with the poisonous yew. See more info in the next section.

Similar species / look-alikes

- **Engelmann spruce** (*Picea engelmannii*) Up to 40 m tall. Very similar to white spruce. Cone scales are flexible, narrow at the tip with a jagged edge. In comparison, white spruce has stiff scales with a smooth rounded tip. Engelmann spruce needles tend to point a bit more upward. Grows at slightly higher elevation, but sometime the two species overlap in a region and hybridize. They then show characteristics between the two species. The hybrid has been given the name "Interior Spruce" (*P. glauca* x *engelmannii*). Both can be used in the same manner as white spruce.
- **Black spruce** (*Picea mariana*) Smaller conifer (up to 15 m tall). Needles are short, stiff, and sharp tipped. They grow radially around the entire twig. Cones are round to oval. The scale edges have tiny teeth. The cones remain on the tree for a long time (some 20 to 30 years). Grows at low to mid-elevation. Edible.
- **Douglas-fir** (*Pseudotsuga menziesii* var. *glauca*) Needles are rounded at tip and flat in cross-section. Bark has deep grooves in orange, yellow, and grey. The cones have three-pronged bracts that hang down below the

Figure 53.8. Douglas-fir cones.

Figure 53.9. Pacific yew.

scales and make them unmistakable. Edible. See Douglas-fir chapter on page 226.

- **Fir** (*Abies* spp.) Bark is smooth with resin blisters on young specimens. Needles are flat, notched at tip. Cones sit upright in tree rather than hanging down; they disintegrate on tree instead of falling to the ground. Edible.
- **Larch** (*Larix* spp.) Deciduous conifer (loses its needles in winter). Needles are held in clusters. Cones have wide scales and bract tips that extend beyond the scales. Edible.
- **Western hemlock** (*Tsuga heterophylla*) Up to 35 m tall. The short needles are evergreen and borne flat on twigs. They are of unequal length and give a feathery impression. Cones are small (1 to 3 cm). They start off purplish green and become light brown when mature. Scales are thin with a wavy edge. Edible. See western hemlock chapter on page 237.
- **Pacific yew** (*Taxus brevifolia*) Small shrub-like tree (up to 15 m tall). Needles are flat, dark green on top and yellow green below. They have pointed tips and are arranged spirally on twigs. Fruit is a fleshy red cup (aril) with a hard seed. Bark is flaky and the underbark is pink/purple. **Very poisonous.**

Western Redcedar

Thuja plicata | Cupressaceae

Figure 54.1. Western redcedar trunk with fibrous bark.

Synonyms

Giant arborvitae, red cedar, British Columbia cedar, British Columbia red cedar, western cedar, giant cedar, Pacific red cedar, shinglewood, western arborvitae, incense cedar

Description

Large evergreen tree (up to 60 m tall, 2.5 m diameter). It often lives up to 1,000 years. **Root system** is shallow but strong and widespread. **Trunk** is tapered, with broad buttresses at base. **Bark** is red/brown and fibrous with narrow longitudinal ridges, soft to the touch. **Leaves** are fern-like and composed of scale-like leaflets that overlap. Leaves lay flat against the **twig**. They are dark green, shiny, and emit a pleasant scent when rubbed. **Flowers** are tiny and cone-like. Male flower is yellowish, female flower is pinkish. They both grow on the same tree. Male flowers have yellow pollen that is spread by the wind. Female flower becomes an erect oval **cone** (1.25 cm long) covered in scales. Each scale has a small prickle on its tip and holds two small winged **seeds**. Cones open at maturity to release the seeds. The empty cones often remain on the tree over winter.

Western redcedar became BC's official tree in 1988. *Thuja* is a Greek word for "trees with aromatic wood"; *plicata* is also Greek, meaning "folded, pleated or folded in plaits," referring to how the leaves are arranged.

Uses

There are a few culinary uses for cedar. I use the leaves for tea or as a trail nibble. Do this sparingly as it contains thujone, which is not recommended in great quantities. See caution box. Both the children and I like the flavor of the leaves. They are very aromatic and strong tasting. A little goes a long way.

Cedar inner bark can be eaten both fresh and dried. Since this is an invasive method of harvesting food, this should only be done cautiously, if at all. If I harvest inner bark from a live standing tree, I make sure never to ring-bark the tree as this will kill it. The tree stands a better chance of being able to heal itself if the wound is smaller. Pathogens or insects can further damage the tree while it is exposed, and this can cause its early demise. In my experience (which is not very extensive with this food), the inner bark is tastier in early spring than at other times of the year. It is also easier to harvest when the sap is starting to run and inner bark sloughs off effortlessly. At other times, it clings to the tree. The flavor of the fresh inner bark is slightly sweet and chalky. It feels like chewing a very tough piece of meat. It takes some jaw muscles and saliva, as well as time, to consume any of the fresh bark on its own. It gives me a satisfying feeling of fullness after I consume it. I imagine that it can be quite constipating in great quantities. One way to harvest larger quantities of the inner bark without damaging standing trees is to collect it when trees have fallen over in windstorms. Just make sure that the tree you are harvesting from is not decayed by fungi or otherwise invaded as the inner bark could then contain pathogens. The inner bark can also be roasted and ground to flour and used in baking.

Figure 54.2. Scale-like leaflets.

Figure 54.3. Male flowers.

Other uses

Western redcedar has been revered everywhere it has grown and has multiple uses as both medicine and material. Ethnobotanist Nancy Turner and the late Mary Thomas have recorded many

Figure 54.4. Inner bark of Western redcedar.

Figure 54.5. Small root.

Figure 54.6. Woody cone.

historical uses for it among the Okanagan-Colville as well as the Secwépemc. It was used for arthritis, dandruff, and as a sweat house tonic. Personally, I like throwing a few branches into a warm bath when I feel achy and blue. It gives a most relaxing feeling.

I make a tincture with the leaves in rubbing alcohol. I use this antibacterial, antifungal rub on areas of the body to stop fungal infections, such as athlete's foot and toenail fungus. It needs to be applied twice a day for a long duration before you notice a marked improvement. I also like to use this rub on a cloth and wipe windowsills and vents to ensure that no mold or fungus can grow there and create unhealthy air in our home. Cedar essential oil is being studied as a cure for sick building syndrome by improving the air quality in a building in a nontoxic safe

Caution

The leaves of cedar contain thujone, which, if consumed in large doses, can cause convulsions and damage to the brain, kidneys, and liver. In alcohol tinctures, the thujone content is stronger so should not be used internally. Avoid tea, tincture, and essential oil if you are pregnant as it may cause miscarriage. Avoid if you are breastfeeding or have kidney problems.

way. Cedar tincture and essential oil should only be used externally. See caution box.

The small roots of the cedar are excellent for baskets and cordage. I use them to stitch together birch bark baskets. On their own, they make sturdy coiled baskets that are waterproof.

Before I harvest the roots, I ask permission from the tree and give the tree an offering. I make sure my headspace is calm and unhurried and that I am in a state of gratefulness. I scratch around in the earth underneath one of the cedars until my fingers find a skinny root that feels sinewy and strong. They have a thin red outer layer and a creamy interior. I follow them along under the ground as far as I can and then pull them up. I make sure not to snap any other roots in the process. The longer root you can get, the better it will be for stitching. I only take a few from each tree, making sure no roots are exposed when I leave. I soak the roots in water for a few hours before using them. This helps them to not snap when bent and under pressure. They dry strong and hard.

Figure 54.7. Nurse log with baby cedar saplings. Note how they resemble moss initially.

The inner bark is/was used by the Secwépemc to weave mats and trays. These mats are/were used to wrap food when pit cooked. They also use the wood to make paddles, spoons, drying racks for food, and fish-spearing poles. The wood is antifungal, rot resistant, and can withstand the weather well. It has been used for fence posts, roof shingles, and other outdoor structures.

Similar species / look-alikes

- **Yellow cedar** (*Chamaecyparis nootkatensis*) More common on the coast. Tip of the tree is very slender and often bent over. Leaflets are more spread out and spiky than Western redcedar. Leaf-covered twigs are four-sided rather than flat. Yellow cedar branches droop downwards while western redcedar branches tend to curve upward in a J-shape. Leaves emit an unpleasant smell when crushed. Cones are small (less than 1 cm). They start off berrylike when young; at maturation, they are woody and have 4 to 6 scales. Cones are rounded and appear popcorn-like when mature. Has many uses.
- **Eastern white cedar** (*Thuja occidentalis*) Very similar tree found in eastern Canada and northeastern US. Smaller tree. Wood is white/yellow when cut; Western redcedar wood has a red hue. Leaflets are more fanned out and each is more rounded. Eastern white cedar cones lack the small prickle on the tip of each scale. Has many uses.
- **Rocky Mountain juniper** (*Juniperus scopulorum*) Small bushy tree. Often has a gnarly appearance. Leaves are blue-green and smaller than cedar. Foliage appears lacy compared to flat for Western redcedar. Female cones look berrylike and are bluish purple with a waxy whitish coating when mature. Edible. See Rocky Mountain juniper chapter on page 189.

Lodgepole Pine

Pinus contorta var. *latifolia* | Pinaceae

Synonyms

Twisted pine, shore pine, contorta pine, black pine, scrub pine, coast pine, beach pine, Yukon pine, mountain pine

Description

Evergreen tree (20 to 25 m tall, rarely 40 m) with a thin and narrow **crown**. Grows in dry montane areas, in subalpine areas, and near the ocean shore. When growing inland, it is straight trunked and in thick stands. When grown by the coast, it has a gnarled and more stunted appearance. The tree I am describing here is the typical lodgepole pine of the Interior. **Trunk** is straight and no larger than 90 cm in diameter. **Bark** is thin and grey in flakes. It cannot survive fires very well. **Branches** are elastic and do not break easily. They are covered in thin grey bark. **Buds** are resinous, sticky, and egg shaped. **Needles** are dark green, shiny, and held in bundles of two. Each needle is pointed, 4 to 8 cm long. They grow around the entire circumference of the branch and stay on the tree on average 4 to 6 years, sometimes as long as 13 years. **Male pollen cones** are reddish green and grow below the new shoots. They grow in abundant clusters. **Female cones** (3 to 7 cm long) have one small prickle on the tip of each cone scale. Cones grow in pairs or threes along the branches on very short stalks. They are slightly lopsided or asymmetrical in shape. Often they are serotinous,

Figure 55.1. A planted stand of uniform lodgepole pine.

Figure 55.2. Needles in bundles of two.

Figure 55.3. Lodgepole pine cone in winter.

meaning that they need fire to open and germinate. They are sealed with resin and will not release their seeds until the resin has been melted by a fire or extreme heat over 113°F (45°C).

Lodgepole pine has a close relationship with the blue-staining slippery Jack fungi (*Suillus tomentosus*). This fungus creates tuberculate ectomycorrhizae around the roots of the lodgepole pine, allowing the tree to access nitrogen in nutrient-poor soils. In exchange, the fungus receives carbohydrates from the tree to make its fruiting body, the mushroom.

Pinus is Latin for "pine"; *contorta* is Latin for "twisted." It could refer to the gnarled look of the trees in windswept areas by the coast. Others suggest that it refers to the way the needles grow twisted on the branch. The name "lodgepole pine" is descriptive of the historical use of the straight trunks as poles for building lodges and homes.

Uses

This tree is a favorite with my kids for making chewing gum. They take a piece of the pitch/resin, choosing the hardened pitch rather than the sticky viscous stuff, place this ball in their mouth and suck on it gently until it is soft and warm. Then they gently move it around in their mouths to soften it without biting down hard. If they do bite down hard, the pitch/resin sticks to their teeth straight away. This has happened on more than one occasion, and it usually results in the necessity of discarding their ruined toothbrush after vigorously trying to remove the pitch from their teeth. Once the resin has been worked gently for an extended period, the gum is ready to be chewed like a commercial gum. It freshens the breath and eases a sore throat. It also tastes good and does not lose its flavor very fast. Bonus, no weird additives such as aspartame are ingested. You can find great quantities of pitch on trees that have been under attack from the pine beetle or have otherwise been injured. The pitch is a defense mechanism by the tree to ward off attackers and to heal a wound. We make sure not to injure the tree but rather collect from a tree that is already oozing pitch. If we are collecting more than just a small chunk,

Figure 55.4. Hardened resin.

Figure 55.5. Harvesting bark off the branches.

we make sure to take from several different trees rather than just one. The more we remove from one tree, the more that tree must produce to replace it. That can be costly to the tree in terms of energy output. Always think about what is best for the tree.

We do not have enough lodgepole pine on the property to harvest any quantities of the nuts. They taste very similar to the pine nuts you can buy, only they are much smaller. One way to get the flavor of the pine nuts, without the effort of finding them or the cost of buying them, is to use a little bit of the bark from the branches. I shave some off and dry it. I grind it fine and add the powder to my pesto, and it gives that lovely pine nut flavor with an added twist of something citrusy.

The needles, which are full of vitamin C, make a refreshing tea. We often nibble on them raw while we are out.

The raw inner bark can be eaten fresh. It is best harvested in early spring. Later in the season, it will have more of a turpentine flavor. It is very fibrous to chew and hard to swallow when raw, but the flavor is nice. I find it easier to consume if it is dried, roasted, and ground into flour. Reports suggest that you mash the inner bark to a pulp and make it into cakes to bake in the oven. I have never tried this method. One lovely way to eat the inner bark is to take a small twig and gnaw off the outer and inner bark like a mouse would. It is nutty, lemony, and surprisingly easy to eat.

I have also seen mention of tapping this tree of sap in early spring. This is something I would like to try next year. A vanillin flavoring is produced as a by-product when lodgepole pine is turned into pulpwood.

Other uses

The needles and the pitch/resin have antiseptic qualities and can help heal wounds and infections. I find that both are great for respiratory complaints. I like making a steam bath with the pitch to inhale when my chest feels tight and my lungs feel raspy. Chewing on the pitch or the buds can help to combat a sore throat. A tea made from the needles with added honey is lovely when you are suffering from a cold.

The pitch has been used as a glue and to waterproof canoes, baskets, and shoes. I find it is an excellent help when starting a fire using wet wood on rainy days. As it melts, the pitch makes a strong flame that dries the other wood out.

The trunks of the trees are so perfectly made for lodge poles with their straightness and uniformity. Their side branches snap off easily to create a smooth finish, free of interfering knots. I lived in a tipi in Sweden for a couple of years and did not have access to lodgepole pine. Instead, I used an assortment of other trees. When I had been lazy and not debarked and shaved my tipi poles perfectly, the rain dripped from those knots straight on my head or in my bed. Having lodgepole pine for tipi poles would have been lovely.

Caution

The wood, sawdust, and resin from lodgepole pine can cause dermatitis.

Similar species / look-alikes

- **Ponderosa pine** (*Pinus ponderosa*) Much larger tree with orange and black bark. Needles come in bundles of three. Edible. See ponderosa pine chapter on page 202.
- **Whitebark pine** (*Pinus albicaulis*) Lives at high elevations. Needles in bundles of five. Cones lack prickles. Decreasing in population and is considered endangered. Do not harvest.
- **Western white pine** (*Pinus monticola*) Tall conifer (30 to 50 m tall) Needles in bundles of five. Cones are long (12 to 25 cm) and lack prickles. Edible.
- **Jack pine** (*Pinus banksiana*) Grows from Nova Scotia to northern BC and in the northern states, sometimes overlaps with lodgepole pine. Lodgepole pine and jack pine can hybridize when grown in the same area. Jack pine has thicker bark with larger scales. Bark is lighter. Cones lack prickles. Can be used in the same way.

Important fact!

Large outbreaks of mountain pine beetle (*Dendroctonus ponderosae*) have devastated great swathes of lodgepole pine forests. Suzanne Simard writes, in her book *Finding the Mother Tree: Discovering the Wisdom of the Forest*, "that in just one decade 18 million hectares of mature pines died from mountain pine beetle attack...making this the largest insect outbreak in recorded history." One report made in 2014 from the University of Alberta states that 28 million hectares have been killed in western North America. To put that in perspective, that is the equivalent of all the forests in the entire country of Sweden!

The mountain pine beetle is a native species. In the past, they lived in balance with the forest and attacked mostly small and weakened individual trees. Most larger trees could withstand their attack. With milder winters and warmer summers, the beetles' population exploded, and they sought out new territory further north and east. It is predicted that a warmer climate will likely result in more and more forest death.

The dead wood from the fallen lodgepole pine is highly flammable and has been and will be dangerous fuel for the more frequent wildfires we can expect in our increasingly dryer climate. The use of herbicides in the forestry industry have killed the aspen groves and other clumps of deciduous trees that historically have been able to dampen a forest fire. Thanks to the higher moisture content and lack of flammable resin, trees like poplars and birches slow a forest fire down.

An interesting and rather hopeful study, published in *Frontiers in Plant Science* by Shiyang Zhao and Nadir Erbilgin, shows that lodgepole pines that have evolved to have larger resin ducts were more likely to survive pine beetle attacks, and if they survive to reproduce, we could see trees that are better able to withstand future attacks. Other reports paint a darker picture where we learn that the pine beetles carry a fungus (*Grosmannia clavigera*), also known as blue stain fungus, that decimates the beneficial *Suillus tomentosus* fungi that assist the trees in their growth. There is concern that regrowth of lodgepole pine might be hindered by the loss of their mycorrhizal partner, blue-staining slippery Jack (*Suillus tomentosus*).

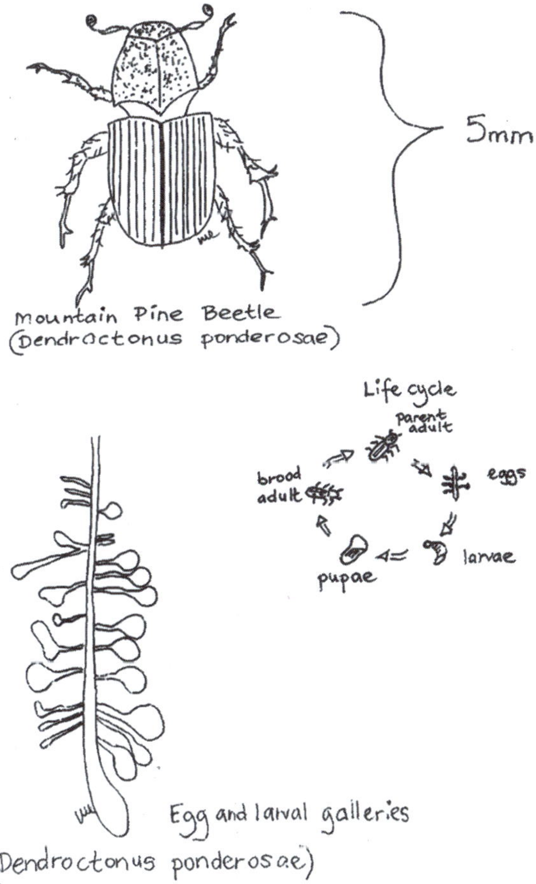

Figure 55.6. Pine beetle, its life cycle and egg/ larval galleries.

Interior Douglas-fir

Pseudotsuga menziesii var. *glauca* | Pinaceae

Synonyms

Blue Douglas-fir, Rocky Mountain Douglas-fir, mountain Douglas-fir, Douglas spruce, Oregon pine, Columbian pine, false hemlock, red fir, red pine, Douglas pine

Figure 56.1. Stand of Douglas-fir on dry hillside.

Description

Large evergreen conifer (up to 40 m) with a diameter of 1 m. Can live up to 400 years. Has a strong, fibrous, and wide-spreading **root system**. Quite drought hardy, will develop deep taproots on dry slopes. **Trunk** is straight and slightly tapered. Has thick and vertically furrowed **bark**; it is corky, grey with yellow grooves, and often has resin blisters. Bark is quite resistant to fire, and the tree loses its lower branches as it ages, further protecting it from ground fires.

Leaves/needles (2 to 3 cm long) are arranged spirally on the branches, held singly rather than in fascicles. They are yellow green and flattened, with a pointed tip, that is not spiky to the touch. Some say the needles smell like camphor when crushed; I have never smelled camphor, so I can not confirm this. **Male flower** is orange red; **female flower** is green with a tint of red, both borne on the same branch. Has very distinct **cones** with three-pronged prominent bracts that stick out past the scales. The pendulous cones are egg shaped (5 to 11 cm long), start off green and later turn brown. They take one season to ripen and fall off the tree when mature. **Seeds** are small and encased in a papery sheath that helps them travel further.

Pseudotsuga means "false tsuga." *Tsuga* is the genus name for hemlocks. *Menziesii* is named in honor of the nineteenth century Scottish botanist Archibald Menzies. *Glauca* means "blue green" or "blue grey." The name "Douglas-fir" is used to differentiate the tree from the true firs, the genus *Abies*. "Douglas" refers to David Douglas, another Scottish botanist, contemporary with Archibald Menzies.

Uses

The needles at this time of year have a high concentration of vitamin C and can be made into tea. When spring comes, the tips of the branches will hold beautiful bright-green soft needles. These are delicious to eat raw, turn into candy (see gummy recipe in spruce chapter, page 213), add to salads and salad dressing, or use in baking. I love chopping them up finely and adding them to lemon-poppy muffins. I replace the poppy seeds with sulphur cinquefoil (*Potentilla recta*) seeds.

Sometimes in winter, the Douglas-fir will drop a lot of seeds on the crusty snow. This is the easiest way to harvest them, easy to spot, and they remain clean. I only collect a few since the seeds are vitally important to squirrels, grouse, mice, and shrews, as well as many overwintering small birds. I grind the seeds (husk and all) until I have a coarse powder, and use it in bread, casseroles, or porridge. It has a slight resinous taste mixed with the sweetness of a pine nut (seed). The seeds are high in protein and contain oils.

I have read about a phenomenon when a crystalline sugar forms on the branches in summer. This sugar is apparently incredibly sweet and was collected when it appeared. The Interior Salish Peoples call it tree-breastmilk. I hope

Figure 56.2. Needles.

Figure 56.3. Male flowers.

to be fortunate enough someday to observe this uncommon event and taste the sugar.

The inner bark/cambium ranges from bitter to sweet depending on when you harvest it; in the spring, it is sweeter. I do not have a lot of experience using the cambium of Douglas-fir. If you harvest inner bark, make sure that you do not take too much and never ringbark a tree as this will kill it.

Figure 56.4. Douglas-fir bleeding with sap.

Figure 56.5. Chewing gum from resin.

Figure 56.6. Cone with bracts sticking out. They look like the two back legs and tail of a mouse.

Other uses

In the middle of winter, I like gathering the small beads of resin that form on the trunks. These small nuggets are perfect as lozenges to soothe a sore throat and can stop a cold in its tracks. I use them as I would commercial throat lozenges.

The needles, made into a tea, can be used to soothe an infected throat and help with a cold. They are slightly diuretic. The needles were used to treat scurvy. They are also antiseptic and can be used to wash wounds.

Douglas-fir wood is highly praised as a commercial lumber species. It is strong and durable. The boughs are comfortable to sit or lie on. We sometimes use them for our forts or when camping.

To help my kids remember which tree is Douglas-fir, I tell them the story about the little mouse who was running away from a fox. It hid in a cone and was safe. To this day, you can still see his back legs and his tail sticking out (the extended bracts). The story is told in many ways, and the author is unknown. I have seen it described as a Pacific Northwest Indigenous legend. This wonderful story connects the kids to the land and helps them to know their neighbors: the trees and animals.

Suzanne Simard has done some interesting research with Douglas-fir and writes beautifully about it in *Finding the Mother Tree: Discovering the Wisdom of the Forest*. She and her team discovered that the Douglas-fir they studied were connected to one another via an underground network of *Rhizopogon* fungus mycelium. The parent trees help their young in life, and even as they die and start to decay, they deliberately direct their last nutrients toward their offspring and other trees in their network.

She also noted that trees of different species helped one another throughout the year. This diversity benefitted all the trees in the forest. Biodiversity contributed to a healthier forest, a forest that could handle stressors better. She states that it is crucial that we understand how important the fungus mycelium is. We need to keep the soil intact in order to have healthy forests that thrive.

Figure 56.7. Pacific yew bark.

Caution

Sensitive individuals might experience skin irritation from the resin. Avoid Douglas-fir (especially the essential oil) if pregnant or breastfeeding.

Similar species / look-alikes

- **English/European yew** (*Taxus baccata*) Small to medium-sized evergreen. Needles are arranged in two flat rows on the branch. Each needle is dark green, flat with a rounded tip with several stomatal bands on the underside. Bark is flaky and brown. Yew has a modified fleshy cone (aril) that looks like a red berry with one seed inside. Take great caution. **Some parts are very poisonous!**
- **Pacific yew** (*Taxus brevifolia*) Similar to English yew. Often a sprawling shrub but can grow larger too. Bark is very distinct, it has flakes in hues of purple, red, or green. When they slough off, you can see the pink underbark. Needles come to a sharp point. They are dark green on top and yellow green below and are held in two rows on the branch. Cone is a fleshy coral-red aril with one seed inside. **Some parts are very poisonous! See more info on page 240.**
- **Western hemlock** (*Tsuga heterophylla*) Evergreen tree with small needles that are borne flat on branches. Needles are unequal

Figure 56.8. Comparison: western hemlock, *top*; Pacific yew, *below*.

Figure 56.9. Larch.

in length. Cones are small (1.3 cm). Edible. See western hemlock chapter on page 237.

- **Fir** (*Abies* spp.) Shares the name "fir" with Douglas-fir, but they are not in the same genus. True firs (*Abies* spp.) have flat needles with a notch at the tip. Cones lack the extended bracts, and they sit upright in tree rather than hang down. Edible.
- **Larch** (*Larix* spp.) Deciduous conifer. Needles are held in clusters. Cones have wide scales and bract tips that extend beyond the scales. Edible.
- **Spruce** (*Picea* spp.) Tall evergreen tree. Bark is covered in thin scales. Needles are 4-sided, not flat. They are spiky to touch; the tip is sharp. I remember which is which by this little saying: "spiky spruce, friendly fir." Cones lack the extended bracts. Edible. See spruce chapter on page 211.

Beaked Hazel

Corylus cornuta | Betulaceae

Synonyms

Beaked hazelnut, California hazel

Description

Deciduous shrub (1 to 4 m tall). Likes clearings, shady openings, and dappled sunlight. Has many **stems** and a cascading look. The slender stems are brown with white lenticels when young, and grey with whitish bloom when older. Grows clumped and spreads sideways by suckers. **Leaves** are oval with double-serrated margins, alternately arranged on the branches. They have a heart-shaped base, with a soft and slightly fuzzy surface. Leaves turn yellow in fall. **Female flowers** are tiny and pink; they look like small anemones. **Male flowers** are formed on drooping catkins. Both types of flowers can be found on the same bush. It flowers early in spring, before the leaves appear, and is pollinated by wind. Once pollinated, the female flower will develop into **husked nuts** that hang in clusters of 2 or 3. **Husk** is papery and covered with small sharp hairs. Inside the husk is a brown hard-shelled **nut**. It is smaller than the hazelnuts/filberts you find in the store. Nut meat is creamy and firm.

Corylus means "helmet," from the Greek *korys*. *Cornuta* means, alternatively, "beaked" or "bearing horns or spurs." Both the genus and species names refer to the husk holding the nut.

Figure 57.1. Beaked hazel bush in summer.

Figure 57.2. Catkins in January.

Figure 57.3. Female flower in early spring.

Figure 57.4. Hazelnut in husk in August.

Uses

The hazel catkins can be eaten in the middle of winter. They are dry and mealy and taste bitter with a hint of something nutty and sweet. We sometimes nibble on them while we are out, but more often we grind them up and add them to our power bars; they are high in protein. Often, we see ruffed grouse (*Bonasa umbellus*) delicately balance its plump body on the slender hazel branches while craning its neck to pick off the catkins.

In winter, when the snow is deep, it can be hard to find plant food in the wild. At this time of year, the larder of food harvested earlier in the year gives us great joy. It is wonderful to reach into the pantry and find dried herbs, seeds, berries, and nuts to add to meals. It brings back memories of happy harvesting moments and gives me a sense of security.

I never find enough beaked hazelnuts to store for the winter though. They are a sought-after food by the animals in the wood, especially the squirrels and bears. By the time they are ripe, they are taken off the tree lickety-split. I do not mind. I can go to a grocery store for

Figure 57.5. Catkin-crumble power bar.

Catkin-crumble Power Bar with Date and Fig Filling (gluten free)

This is a mixture between a power bar and a crumble in texture and taste. It gives energy and is a nice healthy treat. It is best made a day or two in advance and refrigerated until ready to consume. By allowing it a few extra days to settle, it becomes less crumbly and easier to eat as a power bar.

Filling

2 tbsp. ground fresh hazel catkins (I use our magic bullet food-processor, but a coffee grinder or a pestle and mortar could work)
1 cup pitted and chopped dates
⅓ cup water
1 tbsp. coconut butter
3 tbsp. butter
¾ cup chopped figs
10 tbsp. applesauce
3 tbsp. honey

Bottom and topping

2 cups oats (I use old-fashioned oats, but quick oats could work)
2 tbsp. sesame seeds
⅓ cup hemp hearts
1 tbsp. arrowroot
½ cup coconut flour
1 tsp. baking soda
5 tbsp. butter
1 cup brown sugar
1 pinch of salt
white sugar to sprinkle on top

- Mix all filling ingredients and place in a pot. Cook on low heat until the dates and figs are soft, and the mixture can be smeared. It will be very sticky.
- Mix all bottom and topping ingredients together in a bowl. Place half of this mixture in a baking pan. Spread the filling mixture on top of this layer, carefully covering it entirely. My pan was 25 × 25 cm (10 × 10 inch), and that seemed to work okay. If you have a larger pan, you might have to make more of the filling to cover the entire pan. Add the remaining mixture on top of the filling and press down gently. Sprinkle a light layer of white sugar on top.
- Cook for 35 minutes at 400°F (200°C). Allow to cool before eating. It is best after a few days in the fridge. We enjoy them as picnic snacks and as energy boosts for early morning starts. They also taste good with a dollop of thick, creamy Greek yogurt.

Figure 57.6. Nut with husk.

Figure 57.7. Opened nut.

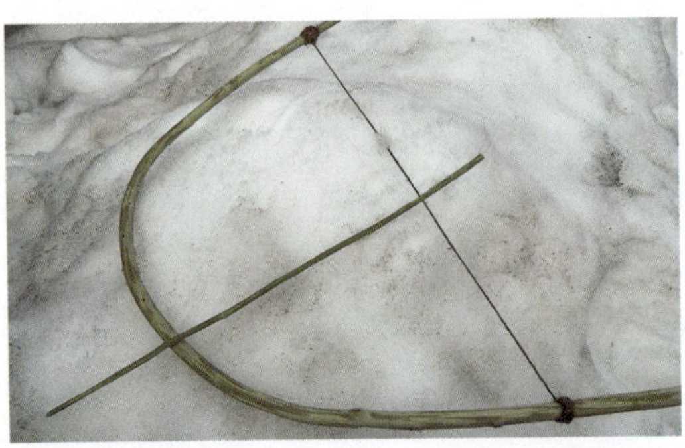

Figure 57.8. Bow and arrow that the kids made from hazel branches.

my food; they cannot. The hazelnuts were traditionally used extensively by both the Syilx and Secwépemc as food and for trading. The nuts are high in magnesium, potassium, phosphorus, and calcium. They are a high source of vitamin E and fatty oils. They are also recommended for pregnant women as they are exceptionally rich in folate (vitamin B9). When I was pregnant, my midwife advised me to consume a certain amount of folates each day to minimize the risk of neural tube defects in the newborn. Personally, I have a slight intolerance to hazelnuts when they are raw, so I avoided them when I was pregnant to be on the safe side. They give me a scratchy throat. Somehow, they do not seem to give me a reaction when they are roasted and coated in dark chocolate. I know...it sounds made-up, but I swear it is true.

Hazelnuts also contain manganese, copper, iron, zinc, and selenium, as well as dietary fiber. Further, they are rich in mono-unsaturated fatty acids like oleic acid, as well as linoleic acid. Oleic acid reduces blood cholesterol levels, while linoleic acid reduces arteriosclerosis and blood lipids and prevents high blood pressure. Hazelnuts can be part of a diet to help prevent coronary artery disease.

Other uses

Nancy Turner writes, in *Plants in British Columbia: Indian Technology*, that the young straight sucker shoots can be peeled and then twisted into a rope to lash things together. She has also recorded the use of the roots and inner bark for making a blue dye. Hazel stems are used to make split hazel baskets, and it is a joy watching people who are experienced weave them. They make it look so easy.

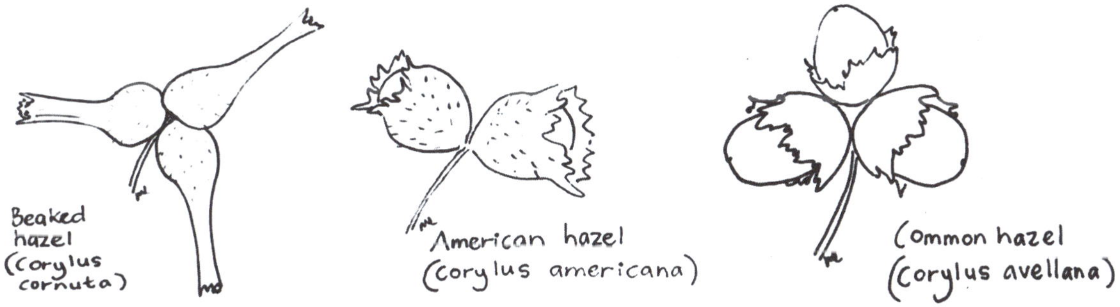

Figure 57.9. Comparison of hazelnuts, left to right: beaked hazel, American hazel, and common hazel.

> **Caution**
>
> Be cautious when picking the nuts as the hairs on the husks easily break off and can lodge in your hand, potentially causing itching and irritation.

Similar species / look-alikes

- **American hazel** (*Corylus americana*) Shrub (2.5 to 5 m tall, 3 to 5 m wide). Very similar. Its nut is entirely encased in a prickly husk, but it lacks the extended spout of beaked hazelnut. See drawing for comparison. Edible.

Figure 57.10. Comparison of leaf and catkin: paper birch, *left*; beaked hazel, *right*.

- **Common hazel/European hazel** (*Corylus avellana*) Shrub (3 to 8 m, can reach 15 m tall). Native to Europe and western Asia. Husk only holds the nut at the base. Husk is velvety and lacks the sharp hairs. Nut is larger. Edible.
- **Paper birch** (*Betula papyrifera*) Tree (up to 20 m). Bark is white with black markings. Leaves are very similar at first glance. Birch leaves have a shiner look and lack the fuzzy texture of the hazel leaves. Does not produce a nut, rather seeds attached in a catkin. Edible. See paper birch chapter on page 242.
- **Witch-hazel** (*Hamamelis* spp.) Shrub (3 to 12 m tall). There are three native species in North America. Female flowers are yellow and bloom early in the spring on a bare twig. Fruit is a capsule in two parts. It has an outer husk that splits and explosively ejects the shiny black seeds at maturity. Can be used externally on skin.

Western Hemlock

Tsuga heterophylla | Pinaceae

Synonyms

Hemlock, British Columbia hemlock, Alaska pine

Description

Large evergreen tree (30 to 50 m tall). Long-lived, some individuals recorded to be 1,200 years old. Tolerates shade and needs a nearby source of water or abundant rainfall. Rarely forms pure stands, but rather mixes with Douglas-fir, spruce, and cedar. Has a **shallow root system**. **Bark** is brown and covered in small scales when young; when old, it thickens and becomes furrowed into flat ridges. Tree has a narrow **crown** with a bent-over drooping tip. I think this makes them look like trees from a Dr. Seuss book. The foliage is feathery and fine. **Needles** (1 to 2 cm) are unequal in length. They are held in two levels on the **twig**; the top level has the shorter needles. Each needle is almost flat with a finely serrated edge and a rounded tip. Top of needle is glossy and dark green; the bottom has two white stomatal bands. Each needle is attached to the branch with a tiny threadlike stem that can be hard to see. Male and female **flowers** are borne on the same tree, but in different locations. Male flowers are formed at the base of the needles, near the tips of a new shoot, starting as pink globes and later turning yellow. Female flowers are scaly, starting off green and later turning pink or purple. They form at the

Figure 58.1. Hemlock tree draped in lung lichen.

Figure 58.2. Underside of needles, cones, and tops of needles.

end of short twigs. Hemlock is wind pollinated. The female flower develops into a small **woody cone** (less than 2.5 cm long). Cones are egg shaped and pendulous. Initially they are green, and at maturity, they turn pale brown. In autumn, the cones open to release small winged **seeds**. The empty cones can remain on the tree over winter.

The name *Tsuga* comes from the Japanese *Tsu-ga*, which combines the elements for "tree" with "mother." *Heterophylla* is Greek and means "different leaves," referring to the two rows of different length needles on each stem. The name "hemlock" confuses the hemlock tree with the herb poison hemlock (*Conium maculatum*), a deadly biennial in the carrot family. They are not related at all. Apparently, the tree was called "hemlock" after the poisonous herb because they smell similar when crushed. I am not sure I agree.

Uses

So far, I have used the winter needles as a spice and as tea. As a spice, I think it tastes like a mixture between pine and dill. The needles are delicious when used to flavor fish. They work very well when finely chopped and sprinkled over a baked potato with sour cream. The needles are high in vitamin C. The tea is mild and pleasant tasting. I remove the needles from the branches before making tea, as I have heard that the tannins in the twigs make the drink bitter. In early spring, the soft tips can be eaten raw.

The inner bark can also be eaten (I have yet to try this), raw, cooked, or dried and ground into flour. If you harvest inner bark, make sure that you do not ringbark the tree, as this could kill it.

Other uses

Hemlock has many recorded medicinal uses. The needles are helpful in a drink to boost the immune system and get through a cold. A needle tea mixed with lemon juice is especially helpful with phlegmy chest complaints. The bark can be used to make skin toners and astringent washes for skin irritation. I have read that the powdered dried bark can be placed in shoes to alleviate foot odor. The sap can be added to rubs for muscle soreness and on the chest while dealing with coughs and heavy lungs.

The bark has been used as a tanning agent and as a pink dye.

In an interesting webinar, I learned that the Sechelt First Nation (Shishalh Nation) use the boughs to collect herring roe. They place them in the water, and the herring come and deposit their eggs on them. Some of those eggs are then pulled up and eaten either fresh or dried. The dried eggs were an important trading item with people inland.

Getting to Know a New Plant

Western hemlock is a new plant for me, one I am just getting to know. I have been learning about, and visiting, the western hemlock for three years now, but do not feel like I know it fully yet. In this chapter, I plan to show you how I get to know a new plant.

The first thing I do is to visit its habitat and walk around in it, noting all the species that share its home. Are there any plants that resemble it? In this case, it was growing side by side with the Pacific yew (*Taxus brevifolia*), a poisonous plant. Once I discovered this, I spent a day sketching, photographing, and taking notes on all the ways you can tell the hemlock apart from the yew. See Similar species/look-alikes section.

By sketching a plant, you hone your observational skills. I note what kind of plants, lichen, moss, and mushrooms this tree collaborates with. Many of the hemlock trees were draped in lung lichen (*Lobaria pulmonaria*). From this fact, I can deduce that the air quality must be good here. The lung lichen only grows where there is unpolluted air, and it is becoming increasingly rare. It is a species that is severely threatened by logging and climate change. Lung lichen needs humid, mature, or old-growth forests to thrive. Seeing it in abundance on the hemlocks at the Kingfisher Interpretive Centre makes me so happy. They are stunning together! The tree seems to do well in the shade, but eventually it will take its place as a towering adult of the forest. Hemlock is considered a climax species, a species that takes its time growing and eventually outlives the pioneer (faster-growing) species around it.

Figure 58.3. Western hemlock bark.

The next step is to read up on the plant. Check out the reference section to find the sources I used when writing this book. The research part of getting to know a plant is not something I rush through; this process can take months. I love this part! You find so many wonderful facts about the tree/plant you are learning about! I try to learn as many synonyms for the plant as I can, and the lore behind how it got its common, Indigenous, and scientific names. There is a wealth of information to be gleaned from them; each tells a story of how people have interacted with that plant. If I can find people in the community who know the plant well, I ask if they would share their knowledge with me.

After doing all these steps, I start to experiment with using the plant.

Caution

Very similar to the poisonous Pacific yew. Check out Similar species/look-alikes section. The tannins in the plant can interfere with absorption of medication and decrease its effectiveness. One report about eastern hemlock states that it is best avoided if you take lithium medication (used in mood stabilizers). Hemlock may reduce the body's ability to rid itself of lithium, resulting in too high levels. Talk to your doctor before mixing the two.

Similar species / look-alikes

- **Eastern hemlock** (*Tsuga canadensis*) Very similar (up to 20 m tall). Found in eastern parts of Canada and the US. Eastern hemlock has larger needles that are flatter on the branch. The top row of western hemlock needles points slightly upward rather than lying completely flat. Can be used the same way.
- **Mountain hemlock** (*Tsuga mertensiana*) Small tree (up to 15 m). Likes higher altitude. Needles grow around the twig as opposed to flat in rows, and they are skinnier. Cone is oval and much larger (can be up to 7 cm long) and pendulous. Can be used the same way.
- **Douglas-fir** (*Pseudotsuga douglasii*) Evergreen tree. Needles attached all around the branch. Each needle is flattened and pointed. Appears more yellow than hemlock. Has a larger cone with three-pronged bracts sticking out. Edible. See Douglas-fir chapter on page 226.

Figure 58.4. Yew aril (berry). Seed is very poisonous!

Figure 58.5. View of top side of needles: hemlock, top; Pacific yew, below. Note the double row of needles on hemlock and single row on yew.

- **Pacific yew** (*Taxus brevifolia*) Shrub-like tree (up to 8 m). Bark is peeling, varies in color from green to dark purple. Needles are spirally held but appear to be in rows. It does not have the two-tiered levels like the western hemlock. Underside of the leaf is pale green yellow, rather than white. Each needle tip is pointed. Branches are often yellow rather than brown. Has a coral-red, fleshy cup-shaped berry (aril) with one large seed instead of a cone. **Poisonous!**

Figure 58.6. View of bottom of needles: hemlock, above; Pacific yew, below. Note how hemlock has white underside while yew is yellow green.

Paper Birch

Betula papyrifera | Betulaceae

Figure 59.1. Paper birch fall colors.

Synonyms

American white birch, canoe birch, mountain paper birch, silver birch

Description

Deciduous tree (up to 25 m tall). A short-lived species that cannot handle extreme heat or humidity very well. Typically lives between 30 and 100 years. **Root system** is rather shallow. **Trunk** (up to 70 cm in diameter) is covered in white flaky **bark** when mature. The peeled-off bark is white with black markings on the outside, and creamy pink on the inside. Young trees have shiny chestnut bark with light-colored lenticels (slits in bark that allow exchange of gases through the bark tissue).

Leaves grow in an alternate pattern and are attached to the **twig** with a rounded petiole (leaf stem). The leaves are oval to triangular, with a pointy tip and a coarse double-toothed margin. In summer, they are dark-green and glossy on top, pale green below and sometimes coated with a downy fuzz. In autumn, they turn yellow. **Male catkins** are formed the previous autumn; the **female catkins** form at the same time as the spring leaves appear. Male and female catkins are borne next to one another, on the same twig. Female catkins are smaller, with greenish flowers, male **flowers** are brownish and are formed on large pendulant catkins. The female **flowers**, when pollinated (by wind), become tiny **two-**

winged nuts/seeds tucked in between the bracts of the female catkin. The female catkin becomes pendulous in fall. The seed-cone/catkin sometimes remain on the tree over winter, but eventually disintegrates and spreads its seeds with the breeze. The tree will start to produce seeds at around 15 years of age. Wood is light-colored and straight grained.

Betula is Latin, originating from an older Gaulish word, *betu*, meaning "bitumen." It was named so because the Gauls extracted tar from the birches and used it as an analgesic and a stimulant. *Papyrifera* means "bearing paper" and refers to the paperlike bark.

Uses

Paper birch is one of those rockstar plants that support humans with food, beverage, medicine, material, and firewood, not to mention aesthetic decoration. In January, we use the catkins as food. At this time of year, the male catkins are the only ones on the tree; the female catkins will form when the leaves do. The catkins have a wintergreen flavor mixed with bitter and sweet.

My mouth gets scratchy if I eat too many raw catkins, but I do not have that reaction when they are roasted, ground, and added to power bars or granola. The catkins can also be steeped for tea.

In February/March, as the snow gives way to muddy spring grounds, the sap starts to run. The best days for tapping sap are when the sun starts to feel warm during the day, but the nights are still below zero. Birch sap is refreshing to drink as a tonic. If you are planning to make syrup, you need a lot of sap: 100 liters of sap to make 1 liter of syrup. I have had birch syrup, and it is truly delicious, but I find it too time-consuming

Figure 59.2. Bark from a rotten log becomes a crown.

Figure 59.3. Underside and topside of leaf.

Figure 59.4. Upright female catkins and hanging-down male catkins.

Figure 59.5. Early spring leaves.

Figure 59.6. Harvesting inner bark.

to make myself. It is filled with minerals and vitamins, such as manganese (very high source, 130 percent of daily value), magnesium, calcium, zinc, folic acid, potassium, phosphorus, and vitamin C.

The sap tastes sweet and is very thirst-quenching. For more info on how to tap sap, see Norway maple chapter on page 11.

The young spring leaves are a vibrant light-green and make a tasty addition to a salad when fresh. I like to dry them too, to use as tea. They are diuretic and cleansing.

The inner bark (cambium) is another fantastic gift from this tree. Found inside the white outer bark, it is orange brown and spongy when fresh. I only use the birch cambium when we have chopped down a tree for firewood. We only chop them down if they are showing a dying top, which will result in them falling in a few years and be a hazard. I do not take cambium from healthy standing trees as this could do unnecessary damage to them. Due to climate change and increasing heat during summers, the birches have had a tough time in this area. They succumb when exposed to extreme heat. The last few years we have had extensive drought and temperatures reaching as high as 115°F (46°C), which is taking its toll on the trees. It is important that we do everything we can to help them thrive; that means not harvesting from healthy trees or clearing them out of the forests. This will give them a greater chance of growing old and be able to repopulate the forest with their offspring.

If you plan to take cambium from a tree that has just fallen, make sure to check for signs of fungi within the inner bark as well as on the outside of the bark. It is hard to identify just what type of fungi has invaded the tree; consuming that fungi may or may not be healthy for you. Fungi can show up as polypores (shelflike mushrooms) attached to the outside of the trunk, or as a discoloration of the cambium. I avoid using cambium from trees with these signs.

To harvest the cambium, I shave off chunks and dry them slightly before roasting in the oven. The last batch I roasted for 15 minutes at 350°F (175°C) and then turned off the oven and left it in there overnight. The next morning, they were nice and crumbly. This crumbly stuff I grind finely into flour. The finer you grind it, the easier it is to bake with. I add this flour to cookies and bread.

Other uses

Here on the farm, many birches have fallen prey to what I believe is the bronze birch borer (*Agrilus anxius*). The name *anxius* means just that: "anxious or troublesome." They attack trees that have been stressed by drought or unusually high temperatures, and as the borers eat their way through the tree, the birch first develops stunted yellow growth, and later complete dieback of upper limbs. A sign that the bronze birch borer is the culprit are D-shaped holes in the bark, their exit holes.

Once the trees develop dead tops, rotten limbs can fall off without warning. They call these trees widow-makers for this reason. We are trying to fell these trees safely before the sap rises too much. This way, the wood (that can still be salvaged for firewood) will season faster. Birch makes the best firewood.

Before we split the logs, I remove the bark to use for crafts, such as basket-making and decorations. I lay a sheet of bark flat, with weights on top to keep it from curling up. Once it has curled up, it is much harder to use. I store it flat, out of the rain, until I want to use it. I am no expert at making birch baskets, but I love learning how to do it. Before using the bark, I soak it in cold water to soften. I scrape the bark with a knife until it has the required thickness. For smaller, more delicate work, I use skinnier sheets of birch; for bigger baskets, a thicker sheet works better. In Sweden, people make long strips of the bark and then weave them into baskets and ornaments. I like making heart-shaped mini baskets with this method.

Adele (Ethel) Thomas from Splatsín te Secwépemc makes beautiful sturdy birch bark baskets with decorative rims. She learnt the art from her Elders and is sharing her skills with the younger generation. She says that it is harder and harder to find good-quality birch bark. It is harder to find the thickness of bark that she was used to in the past. Right now, she uses an artificial sinew to sew the baskets together, but traditionally they were stitched with animal sinew or with cedar or spruce roots. (See chapter on cedar for info on uses of the roots, page 216.) In the past, the Secwépemc used the birch bark baskets to transport and store food and some

Figure 59.7. Paper birch basket by Adele Thomas.

Figure 59.8. Mini canoes as Christmas decorations.

were even used to keep liquid. The bark is waterproof, and the baskets were constructed so well that they did not leak.

The birch bark is also used to make baby baskets and cradles. The baby is comfortably carried in the basket, protected from the weather and in close contact with its caregiver. The insides of the baskets were traditionally filled with soft buckskin or fabric while the bottom contained soft materials such as bark, leaves, and grass that added comfort and wicked away moisture.

The small bits of bark, which cannot be used for basketry or crafts, we put aside to be used for fire starters. They are excellent, making a bright hot flame that can set your kindling alight. Leftover bark can also be made into birch oil and tar. Birch oil can be used as a topical ointment for minor cuts and for poison ivy exposure. It is also a mosquito repellent, as well as a way to protect leather, wood, and metal from rotting and rusting. If the oil is rendered down to a thicker consistency, it can be used as glue and for waterproofing.

Figure 59.9. Birch bark basket sewn together with cedar roots. Red osier dogwood handle and rim.

Chocolate Chip/Birch Bark Cookies

2 eggs	1 tsp. baking powder
1 cup brown sugar	2 tsp. vanilla
2 cups flour	1 cup chocolate chips
¼ cup birch bark flour	½ cup finely chopped dates
1 cup butter	
1 tsp. salt	

- Whip eggs and sugar. Add butter and vanilla.
- Combine all dry ingredients, then slowly mix into egg mixture.
- Gently fold dates and chocolate into the mixture.
- Form dough into balls and place on parchment paper-lined trays. Bake for 10 minutes at 350°F (175°C). Let cool.
- Enjoy!

Figure 59.10. Chocolate chip/birch bark cookies.

Caution

Birch tar can be irritating to the skin. Do not use birch tar or birch oil if you have edema, or poor heart or kidney functions. The leaves are diuretic and if taken with prescribed diuretics can cause the body to lose too much water. Check with your doctor before using. Birch can cause seasonal pollen allergies.

Similar species / look-alikes

- **Trembling aspen** (*Populus tremuloides*) Often mistaken for birch. The bark of aspen does not flake off. It tends to have slightly grey/blue bark while birch bark is brightly white. Aspen bark is coated in a powdery dust. Leaves are held with a flattened petiole rather than a round one, causing them to flutter from side to side in the slightest breeze. Leaves are more rounded, bluish green, and has a less-pronounced serrated margin. Catkins are fuzzy and greyish. Edible. See trembling aspen chapter on page 197.
- **Water birch** (*Betula occidentalis*) Native to western North America. Small deciduous tree (up to 14 m). Grows near waterways. Bark is smooth and dark red brown. It does not flake. I am not personally familiar with this tree/shrub. The information here is from books and online sources.
- **Beaked hazel** (*Corylus cornuta*) Deciduous shrub (up to 5 m tall). Leaves are coated in a downy fuzz. Female flower is pink and small. In fall, it carries husked nuts. Bark does not flake. Edible. See beaked hazel chapter on page 231.

Common Barberry

Berberis vulgaris | Berberidaceae

Synonyms

European barberry, barberry, berberis, pipperidge bush, pepperidge bush, jaundice berry, sow berry, trailing mahonia, woodsour, sour-spine

Description

Deciduous shrub (up to 4 m). Likes partially wooded areas. Introduced species from Europe, North Africa, and the Middle East. **Stems** are grey (sometimes with shredding bark), with spines at each node. **Spines** are double- or triple-branched; they look like barbed wire to me. The stems often give the bush an arching form. Individual stems can live up to 30 years. **Leaves** are held in clusters of 2 to 5. The leaf clusters are attached alternately on the branch. Each leaf is teardrop shaped with a serrated edge. In summer, the leaves are dark green on top and greyish below. In fall, they change color to purple, orange, and red. Where the clusters of leaves meet the stem, there is a three-branched sharp spine. **Flowers** (4 to 8 mm) are yellow with six petals and six stamens. The flowers hang in drooping racemes from the branches in late spring, each inflorescence (cluster of flowers) holds 10 to 20 individual flowers. Bees, wasps, beetles, flies, and ants pollinate them. **Fruit** is an oblong, red berry with 1 to 3 **seeds** inside. Berries remain on the bush throughout winter. Seeds are dispersed by birds and rodents. **Root system** consists of a thick fibrous root crown and far-extending lateral roots (as far as 5 m to the side of the bush). As the roots extend sideways, they send up shoots, and create a new root crown below. If the lateral root (rhizome) is severed, vegetative reproduction occurs (a new individual plant is created). The roots have **yellow inner bark.**

Figure 60.1. Common barberry flowers in May.

Berberis is a Latinized form of the Arabic word for the fruit. *Vulgaris* means "common."

Figure 60.2. Berries in October.

Figure 60.3. Berries still hanging on in February.

Figure 60.4. Barberry spines resemble barbed wire.

Uses

In February, there are few berries to find in the woods, which is why coming across a barberry bush is so exciting. Despite snow and cold weather, the ruby-red berries still cling on. They have a tart taste, a little like cranberries. In late fall, they are juicy, but at this time of year they have dried out on the branch, creating a raisin-like version of themselves. The seeds inside the berry look and taste a bit like grape seeds. We eat the berries as a trail snack in winter. I do the greater bulk of harvesting these berries in late fall while they are still juicy. I dry some, freeze some, and make preserves with some. One of my favorite ways to use the dried berries is to throw a handful into rice. The seeds soften inside the berries during cooking, and they give a rice dish a lovely refreshing taste. We see barberries in traditional Middle Eastern dishes such as Persian rice pilaf, and in some North African couscous recipes.

In Europe, people make preserves from the berries. I harvest the berries for jellies in September or October when they are at their peak.

They make the most delicious jelly, bright red with the right combination of sweet and sour. The berries are high in vitamin C, and they also contain zinc and manganese. Make sure that you have identified the berry correctly. The vine of bittersweet nightshade (*Solanum dulcamara*) likes similar habitats, and its berries are quite similar, red and juicy. They look appetizing and can be enticing for young kids. **Bittersweet nightshade is very poisonous, do not ingest! See similar species/look-alikes.**

In Iran, the berries (known as *zereshk*) are grown commercially, and people use them extensively for juice, marmalade, jelly, jam, pickles, syrups, candy, and liquor.

Other uses

According to an article by Farhood Golmohammadi and Mohammad Karim Motamed in *African Journal of Plant Science*, doctors in Iran suggest drinking barberry juice for treating fever and inflammation. In Iranian traditional medicine, the astringent berries are known for their anti-arrhythmic and sedative effects. The berries were used to decrease blood glucose levels and to lower blood cholesterol. Furthermore, they found that the berries had a strengthening effect on the liver and stomach. They were particularly effective against liver parasites.

Barberry is a relative of Oregon grape, and like its cousin, it also has berberine in the bark of its stems and roots. Berberine is bright yellow and bitter. It is considered to have antimicrobial (killing bacteria and parasites), anti-inflammatory, hypotensive (causing a lowering of blood pressure), sedative, and anticonvulsant effects. See caution box.

Many bird species are attracted to the barberry fruits: ruffed grouse, quail, ring-necked pheasants, cedar waxwings, robins, thrushes, and catbirds are some examples. The fruits/berries are a good source of food in winter or early spring when other sources are scarce. I have seen songbirds nest within the bushes, keeping safe from predation behind a mass of thorns.

Important information

Barberry can be a host to wheat stem rust, which is caused by the fungus *Puccinia graminis tritici*. It causes a disease that affects wheat and sometimes other cereal grains, and it can decimate crops quickly. The fungus has as its host, wheat, and as alternating host, the barberry bush. Barberry has been eradicated in many parts of Canada and the US due to the fear of virulent strains of the disease developing on the bush through genetic recombination. Only stem rust-resistant varieties of barberry can now be imported and sold in Canada and the US.

Growers around the world are breeding for wheat that is more resistant to the rust and try to keep up with various new strains as they appear. Wheat stem rust has been the cause of crop failures and famines in the past. A new virulent strain called Ug99 was found in Uganda in 1998, with recent outbreaks in Scandinavia and Western Europe. The stem rust favors hotter temperatures, and due to climate change and rising temperatures, we are likely to see greater outbreaks of it as summer temperatures arrive earlier. We might also see the spores of the fungus travel further with increasing storms and stronger winds. The rust appears on the barberry leaves as little orange bumps and on wheat as orange lesions, which coat the entire grass stem (culm). To avoid spreading the bush to areas

Caution

Avoid using the root or stem when pregnant or breastfeeding. Do not use for newborns or very small children. Exposure to berberine for a fetus or newborn with jaundice can cause kernicterus (brain damage), which is very rare and preventable. Do not use in cases of jaundice, it can worsen the symptoms. Avoid altogether in cases of liver and kidney problems. Consult with your doctor before using.

near farmlands, harvest it where it grows, but do not transplant. If you want to plant a barberry near your home, use some of the varieties that are resistant to the stem rust.

In the UK, the destruction of barberry bushes has had a devastating effect on the barberry carpet moth (*Pareulype berberata*). It feeds solely on barberry leaves and is now nearing extinction.

Similar species / look-alikes

- **Bittersweet nightshade** (*Solanum dulcamara*) Perennial woody vine. Bittersweet nightshade is in the tomato family, Solanaceae. Flowers are purple five-pointed stars with yellow stamens and style in the center. Leaves are arrow shaped and often lobed at base. Berries are red, juicy, and hang in clusters. Inside of berries are small, yellow, flat, and soft seeds (that look like tomato seeds). **All parts of the plant are poisonous.** Do not ingest. It can cause headache, dizziness, trouble speaking, low body temperature, bleeding in stomach or intestines, trouble breathing, poor blood circulation, convulsions, and possibly even death. Berries look very appetizing, so be aware that small children might be tempted by them. They often grow together with common barberry.

Figure 60.5. Bittersweet nightshade—poisonous!

- **American barberry** (*Berberis canadensis*) Native to eastern North America. Very hard to tell apart from common barberry. I have not seen these in the Okanagan. It has wrongfully been mistaken for common barberry and been eradicated in many places. It is now considered vulnerable to extinction. American barberry is smaller (up to 2 m), and second-year branches are brown/red rather than grey. Its leaves are less serrated, and it carries fewer flowers in each raceme (3 to 12). I do not know if this species is susceptible to wheat stem rust. Edible.

Figure 60.6. Japanese barberry bush.

- **Japanese barberry** (*Berberis thunbergii*) Small deciduous bush (around 1 m tall) with dark purple-red teardrop-shaped leaves. Flowers are yellow. The leaves have a non-serrated edge, and on the underside of the branches hang tiny ovoid, bright-red berries in clusters of three or more. The bush has single thorns alternating up the branch. It is not a host to stem rust. Has edible fruits.

Figure 60.7. American bush cranberry.

- **American bush cranberry/guelder rose** (*Viburnum opulus*) Deciduous bush (up to 5 m). Leaves are three-lobed (look like maple leaves). Flowers are in corymbs with white showy, sterile flowers on the outside and smaller, fertile flowers in the center. Fruit is a round juicy drupe that often remain on the bush over winter. It has one large flattened, hard seed inside. Technically edible in small quantities; it makes me nauseated if I eat more than five raw drupes in one sitting. In large quantities, the berries can cause vomiting and diarrhea.

Rocky Mountain Maple

Acer glabrum | Sapindaceae

Synonyms

Douglas maple, rock maple, Greene's maple, New Mexico maple, Torrey maple, mountain maple, three leaf Rocky Mountain maple, rose-fruit Rocky Mountain maple, Sierra maple, California mountain maple, box maple, dwarf maple

Description

Native, small deciduous tree (up 10 m tall). Grows in open forests and clearings. Prefers to grow in gullies and on seepage sites with adequate moisture. Often has multiple, slender **trunks. Bark** is grey/brown, thin and smooth. New growth is vibrantly red, and spring **buds** are pink/red. **Leaves** are opposite on branch, with 3 to 5 lobes that are coarsely double-toothed. They are dark green on top and whitish green below; in fall, they turn orange/red. **Flowers** are in panicles (5 to 10 in each panicle). Flowers are yellow green and appear at the same time as the leaves. Female and male flowers can be on the same tree but are more commonly found on separate trees. Female flower has a curled pistil in its center, and the male flower has multiple pollen-laden stamens. **Fruit** (samara) has two winged **seeds** that are fused together at a 45° angle.

Rocky Mountain maple is often one of the first to repopulate burn sites and can resprout from its root crown if top is damaged or killed. Its **root system** is shallow and wide spreading.

Acer means "sharp" in Latin and refers to the fact that trees in this genus have hard wood and were used to make spears by the Romans. *Glabrum* means "smooth" or "hairless."

Figure 61.1. Young Rocky Mountain maple leaf.

Figure 61.2. Grey bark with lichen.

Figure 61.3. Samara.

Figure 61.4. Male flower.

Figure 61.5. Female flower.

Figure 61.6. Bud.

Uses

In early spring, we tap this maple for sap. While this tree is smaller than Norway maple, I find its sap sweeter. For info on how to tap for sap, see Norway maple chapter on page 11. We drink the thirst-quenching and delicious liquid straight from the tree, without any preparation. I only keep the tap in the tree for three days before removing it. I always want to make sure that I do not take more sap than the tree can manage to go without. To stop the tree from bleeding, I plug it up with some clean beeswax after removing the tap. This seals the wound effectively.

When the leaves appear in spring, we harvest those to add to salads and pasta sauce. They are astringent and a bit bitter. I like their flavor in small doses. The young samaras can be used for food too. As soon as they become too hard to chew, they are best left alone.

Caution

Partially dried or wilted leaves of this maple might be toxic to horses in high doses. Wilted or partially dried leaves of red maple (*Acer rubrum*) can cause hemolytic anemia (red blood cells are destroyed faster than they can be produced) in horses and ponies and can cause death within twenty-four hours of ingesting. The fall leaves are more poisonous than the summer leaves. Fresh leaves do not seem to affect the horses.

Other uses

The inner bark was used by the Nlaka´pmx and the Secwépemc to make rope and twine. The green wood can be soaked and bent to form snowshoe frames. The hard wood is also used to make sticks, bows, rattles, masks, dip nets, and spoons.

The leaves can be packed around apples and root crops to help preserve them for longer. Rocky Mountain maple is important for watershed cover. It likes to grow in gullies where snowmelt water trickles down from higher altitude. By shading these areas, Rocky Mountain maple keeps the water from evaporating as fast and helps to create a moist and protected environment for many species. It is also an important food for moose, elk, mule deer, and white-tailed deer. The tree makes great nesting and cover sites for many game birds, songbirds, and small mammals. The flowers and nectar attract bees and hummingbirds. Thanks to its ability to come back quickly after a wildfire, it plays an important role in restoring damaged areas.

Similar species / look-alikes

- **Vine maple** (*Acer circinatum*) Large shrub (up to 8 m). Native to coastal areas of southern BC and northwestern states. Not common inland. Leaves are long and broad, with up to 11 lobes. The maple wing keys/samaras are wide apart, at an almost 180° angle. Can be tapped for sap.
- **Box elder maple** (*Acer negundo*) Deciduous tree (up to 25 m). Shoots are green, often with a waxy purple coating. Leaves are pinnately compound with 3 to 7 leaflets. Leaves turn yellow in fall. Samaras are at a 45° angle. Edible.

Figure 61.7. Box elder maple.

- **Norway maple** (*Acer platanoides*) Deciduous tree (30 m tall). Leaves have five lobes. Petioles exude a milky fluid. Samaras are at a 140° angle. Edible. See Norway maple chapter on page 10.
- **Red maple** (*Acer rubrum*) Deciduous tree (up to 38 m). Often planted as a landscape tree. Leaves are 5 to 10 cm long, with 3 to 5 lobes; margins are serrated. They turn a brilliant red in autumn. Flowers are red. Pink samaras hang in pairs at 50° to 60° angle. Can be tapped for sap. **Can be poisonous to horses.** See caution box on page 255.

Sulphur Cinquefoil

Potentilla recta | Rosaceae

Synonyms

Erect cinquefoil, rough-fruited cinquefoil, five fingers

Description

Perennial herb (up to 80 cm tall). Grows in open fields with poor soil. **Leaves** are palmately divided into 5 to 7 deeply toothed and veined leaflets, dark matte green on top and pale green below. Leaves form a basal rosette as well as grow in an alternate pattern up the stalk. At each petiole, there are two jagged stipules attached to the stalk.

Stalk is woody and green/yellow. Leaves and stalk are covered in soft hairs. **Flower** is yellow in two tones, light on petals and dark toward the center. It has five petals, each heart shaped. Flowers grow in flat topped clusters (cymes). **Fruit** is small and strawberry looking, enclosed within the calyx. The numerous **seeds** are dark brown and wedge shaped. **Taproot** is fibrous and woody.

The name "cinquefoil" means either "five leaf" or "five petals." The Latin *Potentilla* means "little powerful one" and alludes to its use as medicine. *Recta* means "erect."

Uses

This species is an introduced variety that can become quite widespread. The leaves, in early spring or when they are immature, can be added to any meal you cook. I do not find them suitable raw in a salad due to their texture and the fine hairs on the leaves. The leaves dry quickly and are easy to crumble and add to bread and muffins. They taste like strawberry leaves (*Fragaria* spp.) or rose leaves (*Rosa* spp.): fresh, sour, and astringent. They are all part of the same family, Rosaceae.

The only time I like eating the leaves raw is on a frosty morning. Coated in frost, they melt in your mouth and give you a lingering fresh taste of summer. When they are eaten like this, the hairs are not bothersome, and no cooking is required.

Figure 62.1. Flowers with heart-shaped petals.

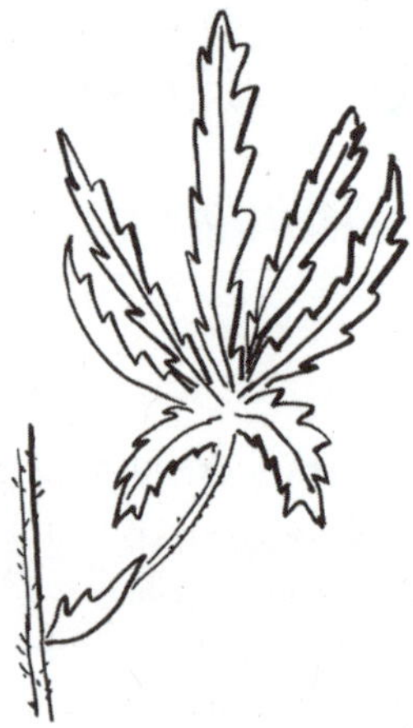

Figure 62.2. Note the jagged stipule next to stalk.

Figure 62.3. Palmately divided leaf.

We also use the flower petals as garnish. They must be added at the very last minute before serving because the petals are fragile and do not hold up to a lot of handling. They look like beautiful little yellow hearts. The fruit is small and enclosed in the calyx. It tastes like an unripe strawberry. Once the calyx dries, the seeds from the fruit scatter freely and abundantly. I pick the plant just before it has released all its seeds and put it upside-down in a paper bag to gather the seeds, which taste a lot like strawberry seeds. The seeds are tiny, but there is usually a great abundance of them since the plant tends to grow in large patches. I find them worthwhile to gather. I use them in the same way as you would poppy seeds.

Other uses

The leaves can also be used medicinally whenever an astringent is called for. The astringency in the plant helps to tighten the skin and heal wounds. I have used it with great success when I have bit the inside of my cheek and the puncture wound has become infected. I make a decoction of the fresh plant, stalk, leaves, and

Figure 62.4. Frozen treat.

Figure 62.5. Sulphur cinquefoil seeds.

Caution

Note that there are two species with the same common name and almost identical Latin names: **Sulphur cinquefoil/Erect cinquefoil** (*Potentilla recta*) that we speak of here (an introduced species from Eurasia), and **Tormentil/Erect cinquefoil** (*P. erecta*) that grows in Europe. Tormentil cannot be eaten as food but has been used for medicine. Any amount over one gram can cause gastrointestinal problems and issues with absorption of minerals and iron in diet. I have not seen tormentil here, but it may be around. An easy way to tell them apart is by the flower petals. Sulphur cinquefoil has five petals, tormentil has four.

flowers. I gargle with this liquid three times per day until I feel better. After using it for only a day, I can already see a marked improvement, and the pain is almost all gone. A tea made from the leaves can also be drunk if you are suffering from diarrhea.

Figure 62.6. Sticky cinquefoil. Note the rounded rather than heart-shaped petals.

Figure 62.7. Buttercup.

Similar species / look-alikes

- **Diverse-leaved cinquefoil** (*Potentilla diversifolia*) Differs from sulphur cinquefoil in that the flowers are darker yellow and lack the two tones. Has heart-shaped flower petals too and palmate leaves with 5 to 7 leaflets. Mostly basal. Edible.
- **Sticky cinquefoil** (*Potentilla glandulosa*) Has sticky glands. Flower petals are rounded rather than heart shaped. Sepals are very prominent. Edible.
- **Slender cinquefoil** (*Potentilla gracilis*) Native. Flowers are more uniformly yellow and showier than sulphur cinquefoil. Leaves are more finely divided and form more of a basal rosette. Can be used the same way.
- **Buttercup** (*Ranunculus* spp.) Leaves are divided into three deeply toothed leaflets. Flowers have five shiny and bright yellow petals. Used medicinally. **Some species can be poisonous.**
- **Tormentil/Erect cinquefoil** (*Potentilla erecta*) **See caution box.**

Black Cottonwood

Populus balsamifera ssp. *trichocarpa* | Salicaceae

Figure 63.1. Black cottonwood is an important riparian species.

Synonyms

Balsam cottonwood, balm cottonwood, western balm cottonwood, western balsam poplar, California poplar

Description

Fast-growing poplar (up to 40 m tall and 1 to 2 m in diameter). Rarely lives longer than 200 years. Likes moist soil and is intolerant of shade. Often seen by rivers and lakes. **Trunk** is straight; bark is grey and deeply furrowed on old trees. Young trees have smooth green-yellow **bark**. **Twigs** have long, pointed, and slightly curved **buds** that exude a sticky resin in early spring. The resinous buds have a strong, sweet-balsam smell. The oval **leaves** (7 to 12 cm long) have a jagged margin and are held alternately on the branches. Leaves are dark green above and silvery-green below. The plant is dioecious. Pink/red male **flowers** are held in drooping catkins. Female catkins look like chains of bead-like seedpods. Each **seedpod** is divided into three. When ripe, the seedpod bursts open and releases **seeds** in cottony fluff that carries them on the wind. Seeds need a moist seedbed to be able to germinate.

Black cottonwood has an extensive and strong **root system.** It also has high levels of rooting hormone and can easily resprout. This tree can abscise (intentionally drop) green shoots that can take hold, develop roots, and

Figure 63.2. Leaf top and bottom.

grow a new tree. The green abscised shoots can survive travel via waterways and take hold further downstream, and in this way colonize exposed sandbars.

The name *Populus* is Latin for "people," while *trichocarpa* is Greek for "hairy fruit."

Uses

The inner bark can be harvested as a food source in spring. It is used fresh or dried and cannot be stored for very long. Due to its sweetness, it ferments fast and should be used quickly before it turns sour. The catkins can be eaten, but they are very bitter. Some sources talk about tapping the tree for sap in early spring. I have tried but have been unsuccessful. I either did not make the tap deep enough, or I missed the window when the sap was running.

Other uses

In early spring, the poplar buds become sticky and smell sweet like honey. Before they break open into leaves, I harvest these and put them

Figure 63.3. Cottonwoods make great homes for many critters.

Figure 63.4. Sticky spring buds.

Figure 63.5. Seedpod splits in three.

Figure 63.6. Catkin with cottony fluff and seeds.

into a jar with olive oil. Over time they infuse the oil with their wonderful qualities. The oil can be used on its own or added to salves. The buds are an expectorant and work a treat for any chest ailments, such as a persistent cough or wheezy breathing. I smear the infused oil, several times a day, on the chest and back, as well as under the nostrils. When I had chest pains and labored breathing after chronic COVID, I noticed how my symptoms eased when I applied this oil. The buds can be taken internally; herbalists recommend them for bronchitis and upper respiratory tract infections.

The cottonwood bud oil can also be used on skin ailments and for dry, flaky skin. The oil feels nice rubbed on achy joints or on muscles that are painful from overexertion. Both the buds and bark contain salicin, which takes away pain and lowers a fever. They should not be used by people who are allergic to aspirin or salicylic acid. See caution box.

According to Nancy Turner and co-authors of *Ethnobotany of the Okanagan-Colville Indians of British Columbia and Washington State*, large cottonwood trees were used to make dugout canoes that were very light. The resin in the buds can be used as glue for making canoes, spears, and fishhooks. Upper branches were dried over summer and used to make fire drills. The ashes from the burnt tree were used as soap to clean buckskin and to wash the hair. Cottonwood can also be used when smoking hides.

Cottonwoods are important for numerous species of birds as nesting sites and as food for deer, elk, and hare. It is a favorite with the beavers. The trees protect the riverbanks and make a shaded and nutrient-dense habitat for fish. They are an important riparian species.

A conversation with Dave Derbowka about poplars

In Armstrong/Spallumcheen, Dave Derbowka has created a phytoremediation plantation of hybrid poplars. He decided to embark on this project when he saw water from the landfill seeping into adjacent properties. Dave has a background in water treatment and management, and he saw an urgent need to protect the groundwater of the community. He wanted to make sure that water from the landfill was treated and purified somehow, so he started the landfill phytoremediation plantation. The poplars help to remediate the landfill leakage. The water in the landfill gets routed to a pond and then pumped to the trees. As the trees absorb this water, they purify the ground around them. After some years, new trees are planted, and the old ones are turned into biochar. The biochar can be used to fertilize the soil and to filter and purify water. Dave also makes building bricks with it.

The well-being of the planet is foremost on Dave's mind as he speaks about how effective these trees can be for carbon sequestration. They become big trees quickly. As I scan the rows of six-year-old trees, I can see the truth in this. They tower above me and shade the sun, lowering the temperature by several degrees. Another bonus to planting these trees is that due to their high moisture content they can function as effective firewalls in the landscape. A fire will move through a stand of these trees much slower and with lower intensity than through a stand of coniferous trees with highly flammable resin. The poplars that Dave grows have been used for carbon offset. Individuals and townships can pay for trees like these to be planted to lower their carbon footprint. Obviously, the best course of action is to stop emitting so much carbon *and* planting trees, rather than planting trees and carrying on as usual, but any step taken in this direction must be applauded.

Currently there is some exciting research happening at Dave's phytoremediation site. Scientists from various universities are looking into the symbiotic relationship that poplars have with endophytes, fungi that have been around for millions of years. It appears as if the endophytes have a large role to play in how well a poplar copes with polluted lands and how well they can help to clean up the mess that humans have caused. According to the US National Institute of Environmental Health Sciences, poplars that were inoculated with certain endophytes in areas with polluted water were able to thrive. Without the inoculation, the trees became stunted. Dave has been a member of the Poplar and Willow Council of Canada (PWCC) since 1996. PWCC focuses on how to preserve and wisely use Canada's private and public poplar and willow resources.

Figure 63.7. Dave Derbowka at the phytoremediation site.

Caution

If you have an allergy to aspirin, do not use cottonwood bark or buds. If you are pregnant or nursing, check with your doctor before using. Do not give cottonwood bark or buds to children under sixteen when they have a flu, chickenpox, or viral infection. They can develop Reye's syndrome, which is a very serious, albeit rare disease.

Similar species / look-alikes

- **Balsam poplar** (*Populus balsamifera* ssp. *balsamifera*) Grows further north. Leaves are wider. The easiest way to tell them apart is by the seedpod/fruit capsule. The balsam poplar seedpod splits in two when ripe, while black cottonwood splits in three. Can be used similarly.
- **Narrowleaf cottonwood** (*Populus angustifolia*) Smaller tree (up to 18 m). Native to western North America, more common in Alberta than in BC. Leaves are skinny and resemble willow leaves. Can be used similarly.
- **Eastern cottonwood** (*Populus deltoides*) Has been planted in western Canada but is originally a native species elsewhere in Canada and the US. Leaves are triangular, coarse toothed with rounded teeth. Can be used similarly.
- **Trembling aspen** (*Populus tremuloides*) Deciduous tree (25 m tall). Has blue-grey, sometimes white bark, with black markings. Bark is coated in a powdery dust. Leaves are rounded with a flattened petiole. Catkins are fuzzy and grey. Edible. See trembling aspen chapter on page 197.

Lamb's Ears

Stachys byzantina | Lamiaceae

Synonyms

Woolly hedgenettle, woolly betony, woolly woundwort

Description

Perennial (up to 1 m). Native to the Middle East. Garden ornamental that has occasionally naturalized in the wild. Basal **leaves** are thick and large (5 to 10 cm long). They are oval with a pointy tip. Leaf margins are finely crenate. The entire plant is coated in woolly white hairs. **Stem** is square and erect. Leaves on stalk are arranged in an opposite decussate manner (pairs of leaves that rotate 90° at each node). **Flowers** are purple/pink and have a top petal and a bottom-lip petal with rounded lobes. Flowers are attached in rings around the top flower spike and lack pedicels. Under each flower ring there are a pair of leaves, which become smaller and smaller the higher up the flower spike they go. **Seeds** are round and brown. The plant **smells** sage-like, or in my opinion, a bit like a sheep. Lamb's ears have **creeping root stems** that root as they trail along the ground. It can become a dense mat of foliage this way.

Stachys means "an ear of grain" and refers to the flower spike. *Byzantina* means "of Byzantium, Constantinople" (Istanbul, Turkey).

Uses

In February, as the snow melts, I can sometimes find intact leaves of lamb's ears. As it starts to warm up, new young leaves emerge, which can be steamed, stir-fried, and eaten raw in salads. My favorite way to cook these is to pan-fry them. The recipe below is inspired by a Brazilian recipe called *Peixinhos-da-Horta*, which means "little fishes from the garden." They look a lot like battered sole fillets when cooked. They have a crunchy satisfying taste, strangely reminiscent of battered fish. They taste great with tartar sauce.

Figure 64.1. Lamb's ears basal rosette.

Figure 64.2. Flower.

Figure 64.3. Flower spike, note arrangement of leaves and the square stem.

Lamb's Ears Peixinhos-da-Horta

10 large lamb's ear leaves
1 cup flour of choice
2 eggs
pinch of salt
pepper
oil for frying

Dipping sauce
1 to 2 cloves of garlic
chili flakes to taste
½ cup mayonnaise
4 tbsp. sour cream
salt
pepper
parsley, chopped finely
1 fresh lime

- Wash and pat dry the leaves.
- Mix eggs, salt, and pepper. Dip the leaves in the egg batter.
- Spread flour on tray. Place battered leaves in flour and coat them completely on both sides.
- Fry in oil until they are a golden brown on both sides. Serve with dipping sauce.
- To make sauce, combine garlic, chili, parsley, salt and pepper to taste with the mayonnaise. Add sour cream and stir.
- You can choose to squeeze the lime juice directly onto the battered lamb's ear leaves at the table or mix it into the dipping sauce. Either way works.

Other uses

Another common name for this plant is woolly woundwort, and it has been used as a bandage to slow bleeding and help heal wounds. The leaves have astringent, antibacterial, and antiseptic properties. They can be dried and used for infusions for colds and sore throats. Lamb's ears have also been used to relieve some of the symptoms of asthma. I have not used it for that purpose myself, so I cannot say how effective it is.

This plant is a wonderful groundcover and border plant. The deer do not seem to like it; they leave it alone. Lamb's ears foliage is very attractive and excellent for small children, as it is soft and safe to play with. The leaves can be used as toilet paper. The flowers attract bees, bumblebees, and hummingbirds. In early morning, the leaves collect dew and are a source of water for insects on hot summer days.

Caution

Can be invasive.

Figure 64.4. Battered lamb's ears and fried lobster mushrooms, lamb's quarters bread, and chickweed salad.

Similar species / look-alikes

- **Common sage** (*Salvia officinalis*) Perennial herb. Has woody stems. Leaves are soft, but not as woolly as lamb's ears. The veins on the leaves are clearly visible, and the tip is rounded rather than pointy. Flowers can be white, pink, or purple. Edible.
- **Silver sage** (*Salvia argentea*) Decorative landscaping plant from Europe. Leaves are woolly and silvery with large, visibly margined teeth. Flowers are white, with yellow or pink splotches. Some sources state that it is not edible, while others mention that the flowers are edible. I have not tried this myself.
- **Mullein** (*Verbascum thapsus*) Introduced biennial plant, (up to 2 m tall). Large soft, downy leaves are more yellow green (lamb's ear leaves are more silver blue). Flowers are yellow with five petals and are borne on a spike. Has many medicinal uses.

Figure 64.5. Comparison: lamb's ears (two leaves on left), mullein (two leaves on right).

Acknowledgments

Being able to write this book has been such a wonderful journey. I could not have done it without the incredible support from my mom, Joanne Cannon. She knows my fear of computers and all things technical, and she took on the task of taking my writing, drawings, and photos and put them together into a finished draft. She also edited the work, cared for my kids, and kept me on track. This book would not exist today if it wasn't for you, mom! I don't have enough words to express my gratitude for you.

There are so many people who have helped in so many ways. My heartfelt gratitude to all of you! Indra McMorran for her artistic eye and skill taking photos of our family foraging. I want to especially acknowledge and thank Crystal Dawn Morris and Adele Thomas for their patient and loving guidance teaching me about the Secwépemc history. A huge thank you to Nancy Turner for her encouragment and support and for taking the time to proofread the manuscript and helping to improve it. Thanks to Renee Crawford who helped me reach out to publishers and guided me through the process of promoting this book; Carolyn and Kassie Ferguson for the many hours you helped edit and spell-check the document; Adam for lending your eye to the design and format of the book and giving me your sage advice. Thanks, Ingvar Larsson (Dad), for your encouragement (and patience) of my self-sufficiency projects while in your house and for our conversations about all things nature. Thank you Robert Bigsby for letting us live on your land and for telling us stories of what the land was like when you grew up here. We learn so much from you!

To every one of you who read and gave feedback on earlier copies: You are brilliant! Thank you! I also want to take the time to show my deep appreciation for the people who partook in this book and who shared their knowledge: Laura Dohla, Terryn Corbett, Anneh Kessels, David Derbowka, Adele (Ethel) Thomas, Crystal Dawn Morris, Kate Wamboldt, Jami Harding, Eli Pivnick, and Kyle Runzer. You have enriched my life!

I am truly blessed to have been able to work with New Society Publishers; their team worked so hard on this book. Every person from copy editor, to marketing, to layout, to cover design to accounting have been so professional, kind, and talented. It is a dream come true to work with you! I am so inspired by New Society Publishers for their mission to create a better world through literature that inspire and teach. It is an honor to be part of this vision. I want to especially say thank you to Caylie Graham for championing this book and for guiding me gently on the publishing journey. You have been such a joy to connect with, and I hope to go on many foraging excursions with you in the future.

Thanks to all of you who come out to my classes on foraging; you inspire me and teach me in so many ways. I have so much gratitude for the wonderful teachers I have had over the years, in person, in books, or online. Thank you for sharing your wisdom and skills. Special mention to Mary Thomas, Nancy Turner, Samuel Thayer, Ashley Adamant at Practical Self Reliance, Marianne Ignace and Ronald E. Ignace, Suzanne Simard, Robin Wall Kimmerer, Bernice Jensen, Marilyn Alexis, Fiona Hamersley-Chambers, Beryl Amaron, Laila Spik, Laureen Felix, Edna Felix, Gloria Morgan, Elizabeth Jean Brown, Jean Brown, all the teachers at Pacific Rim College, and Bo Weslien and Katha Blank at Sjöviks Folkhögskola. There are so many more who have influenced me and taught me, I wish I could list you all.

Kukstsamc and Limlimt to the Secwépemc and Syilx-Okanagan Nations for your wisdom, land stewardship and kindness.

I am truly grateful to the Splatsin community for sharing their stories and wealth of knowledge with me. My life is so much richer learning from you.

Thank you to my family. You are my inspiration. Thank you, bright, strong daughters, Willow and Phoenix, for who you are, for showing me what is important, and for your exuberant, beautiful energy. It is a joy to be around you, I learn so much from you every day!

Thank you, wonderful life partner, Peter, you are my rock! Thank you for being my steady beacon and for raising our family, bringing music to my soul, and living this magical life with me. I am forever grateful to you.

Finally, I would like to thank the land with my entire heart. I have so much gratitude for all the plants and mushrooms who have shown me so much. You are the reason this book exists. I am deeply indebted to you for food, material, medicine…my life. Thank you for your guidance and abundant nurturing.

Glossary

References to figures in text, see pages 279-287

abscise: (botany) when branch or limb deliberately falls off

achene: fruit with one seed; usually small and does not open to release its seed (fig. 65.12)

acidic: having a pH below 7

adventitious: root that is growing from an unusual place: directly from a stem, leaf, or old root

aggregate fruit: fruit formed when there are multiple pistils in the same flower, e.g., raspberry

alkaline: having a pH greater than 7

allelopathic: when chemical compounds are made by one plant to deter the growth of plants around it

allopathic: term used to describe modern or mainstream medicine

alternate: leaves or shoots that are placed singly at different heights on a stem (fig. 65.10)

annual: plant that lives for a year or less

anther: male flower's reproductive system where pollen is held (fig. 65.1)

arcuate: arching or curved like a bow

aril: specialized outgrowth on a seed that partly or completely covers the seed (fig. 65.12)

asci: (ascus, singular) a sac, often cylindrical in shape; where the spores of ascomycete fungi develop (fig 65.9)

ascomycete: fungi whose spores develop within an ascus

auricle: small ear-shaped lobe or appendage at base of leaf or petal

axil: angle between a leaf stalk or branch and the stem or trunk from which it is growing (fig. 65.10)

banner petal: the large lobed petal at the top of a flower in the pea family (fig. 65.3)

barb: a short stiff bristle or hairlike spine

basal rosette: leaves of a plant arranged in a circular pattern at the base of plant, close to the ground (fig. 65.5)

biennial: plant that completes its life cycle in two growing seasons; flowers and make seeds in its second year

bifurcated: split into two parts

bioluminescence: light produced by a chemical reaction within a living organism

bioremediation: using living organisms such as mushrooms to remove contaminants and pollutants from soil or water

bipinnate: see pinnately compound leaf (fig. 65.7)

bract: modified small leaflike structure that is often directly below the flower or inflorescence (fig. 65.1)

burr: seed or dry fruit that has hooks or teeth (fig. 65.12)

buttress roots: large wide roots that prevent a shallow-rooted tree from falling over

C4 **photosynthetic pathway**: a way for plants to minimize photorespiration and conserve water; commonly seen in plants in warm climates

calyx: name of all the sepals of a flower collectively as one structure (fig. 65.1)

CAM photosynthetic pathway: seen in plants that are adapted to perform photosynthesis under stress; reduces photorespiration. In CAM plants the stomata are open at night and then absorb carbon dioxide and thus reduce water loss during the daytime.

cambium: layer between the xylem (wood) and the phloem (bast) of a stem, root, or trunk; the area of a tree that increases in girth

cane: refers to a hollow, slender, and flexible jointed stem like a reed; can also refer to a slender woody stem that usually grows directly from the ground, e.g., rose or raspberry

cap (mushroom): top part of a mushroom; also called pileus (fig. 65.8)

capsule: dry fruit that opens when ripe (fig. 65.12)

carpel: synonym for pistil; female reproductive organ of a flower; consists of an ovary, a stigma, and usually a style (fig. 65.1)

catkin: cylindrical cluster of single-sex flowers bearing scaly bracts and usually having inconspicuous petals or lacking them altogether; most often pollinated by wind, sometimes by insects; may be erect or pendulous (fig. 65.12)

cauline leaves: leaves that are attached above ground to the main or central stem (fig. 65.5)

chaff: husks of seed

chitin: amino-polysaccharide polymer occurring in nature; give crustaceans, insects, and fungi structural strength in their exoskeletons and cell walls

chlorophyll: green pigment that plants use to make food during photosynthesis

cleistogamy: type of self-pollination where plants choose to propagate by using non-opening self-pollinating flowers when the need arises

climax species: plants that are the last stage in the succession of an area; take a long time to mature and tend to live long and dominate the landscape, replacing the pioneer species that came before them

composite flower: composite flowers have two types of flowers, ray florets and disk florets (fig. 65.3)

compound corymbs: botanical term to describe a flower cluster that has several corymbs attached together (fig. 65.11)

compound leaf: leaf that is divided by a midrib with leaflets coming off that axis (fig. 65.4)

compound raceme: flower cluster with several racemes (fig. 65.11)

cone: seed-bearing organ of evergreens; usually woody, oval or round with scales and bracts arranged around a central axis (fig. 65.12)

conifer: group of plants (mostly trees) that most often are evergreen and carry cones. Some examples of conifers are: fir, cedars, spruce, larch, and pine

corymb: flower cluster where all the flower stalks grow to terminate at the same level, making a flattish top (fig. 65.11)

cotyledon: first leaves to appear from a germinating seed; dicots/eudicots have two cotyledons, monocots have one (fig. 65.2)

crenate: having a round toothed or scalloped edge (fig. 65.4)

cultivar: plant variety that has been created by humans in cultivation through selective breeding; designated in the style *Sambucus nigra* 'Black Lace', 'Black Lace' stating the cultivar

cure (nuts): drying nuts in a well-ventilated, dry, and dark place for a few weeks to enhance their flavor and storage capabilities

cutting: using a section of a plant for vegetative propagation

cyme: group of flowers where each stalk produces a flower; new flowers form from side shoots, and the oldest flowers are on top in the middle; like a corymb, but differs in that the oldest flower is in the middle; in a corymb, the oldest flowers are on the outside (fig. 65.11)

deciduous: (of a tree or shrub) dropping its leaves annually

decumbent: (of a plant or part of a plant) lying along the ground or along a surface, with the end of plant or plant part curving upward

decurrent gills: when gills extend down the stem (fig 65.8)

disk florets: number of small tubular and usually fertile florets that form the disk of a composite flower (fig. 65.3)

dioecious: having the male and female reproductive organs on separate individuals

drupaceous nut: type of fruit in which an outer fleshy part surrounds a shell with a seed inside, for example, almonds and walnuts

drupe: fleshy fruit with a thin skin and a central shell containing the seed, e.g., plum or cherry (fig. 65.12)

drupelet: small individual drupes forming a fleshy aggregate fruit such as a blackberry or raspberry. Each individual drupe in such a fruit/berry is called a drupelet.

ectomycorrhiza: symbiotic association between fungi and the feeder roots of larger plants in which both the plant and the fungi benefit; sometimes the association between the two is crucial for the survival of one or both species

elaiosome: fleshy structures that are attached to the seeds of some plant species; contain lipids and protein; insects (often ants) are attracted to the elaisomes for the food they provide for their young

endophyte: organisms, can be fungi and/or bacteria, that live between living plant cells; form symbiotic or pathogenic relationships with their host

entire: leaf margins not toothed or divided in any way (fig. 65.4)

epicalyx bractlets: an additional whorl around the sepals of a flower; bractlets are small modified leaves around the reproductive structure of a flower (fig 65.22)

epithet: For example, in *Homo sapiens* (scientific name for humans), the genus name is *Homo* and the specific epithet is *sapiens*. It is always written in lower case and italics.

essential oil: concentrated liquid made from plants that has volatile (easily evaporated at normal temperatures) chemical compounds

evergreen: plant that keeps its leaves all year

false gills: see gills, false

fascicle: (botany) a bundle of needles, leaves or flowers growing crowded together, enclosed at the base by a sheath (fig. 65.2)

fascicle sheath: papery sheath at base of a bundle of fascicles, e.g., pine needles; may or may not be deciduous (fig. 65.2)

fibrous root: root that has no prominent central axis and that branches in all directions (fig. 65.6)

filament: slender stalk that holds the anther (and pollen sacs) (fig. 65.1)

floral bracts: see bract

floricanes: second-year canes that produce the flowers and berries (in most varieties of blackberries and raspberries). The first-year canes are called primocanes and only have leaves, not flowers or berries.

flower spike: group of stalk-less flowers growing on an upright stem (fig. 65.11)

fungi: (fungus, singular; fungi, plural) eukaryotic organism; include microorganisms such as yeasts and molds, as well as the more familiar mushrooms

gasteroid fungi: fungus that produces spores inside its basidiocarp (fruiting body) rather than on its outer surface, for example, puffball (fig. 65.8)

genus: (genera, plural) biological classification ranking between family and species. Example: *Homo* is the genus that we share with our extinct ancestors such as *Homo erectus*. *Homo sapiens* is our own species, humankind.

gills: radiating, gill-shaped plates (that hold the spores) under the cap of some mushroom (figs. 65.8 and 65.9)

gills, false: forked ridges or folds on the underside of cap and down the stalk of some mushrooms; attached to the stem rather than detached, and to remove them you must break the stem (fig. 65.9)

gills, short: sometimes the gills do not run all the way from the stem to the cap but only partway. These are called short gills or secondary gills, and may be helpful in identification of a mushroom. (fig. 65.9)

glands: in plants, structures that secrete products such as: resin, oil, enzymes, and salt

glandular: full of glands; glandulose (in texture), lumpy

gleba: fleshy spore-bearing inner mass of fruiting bodies of certain fungi, example puffballs (fig. 65.8)

glomerules: cluster of flowers, formed by densification of a cyme
green manure: also called a cover crop; sown crop that is allowed to wither and be turned over into the soil to help improve its nutrient content
habitat: place where an organism makes its home; has all the environmental conditions a specific being needs to survive
hastate: (leaf shape) having a narrow triangular shape like the tip of a spear (fig. 65.7)
hybrid: offspring from two animals or plants that have different subspecies, breeds, varieties, species, or genera
hybridize: when two different species or variants crossbreed and create a new species or variant
hyperaccumulator: plant that is capable of growing in soil or water with very high concentrations of metals; absorbs these metals through its roots, and can store high levels of them in its tissues
hypha: (hyphae, plural) long branching, filamentous structure of a fungus from its base. In most fungi, hyphae are the main mode of vegetative growth, often underground, and collectively called mycelium (fig. 65.8)
inflorescence: group or cluster of flowers, stems, stalks, and bracts; collective name of this structure (fig. 65.5)
involucre: one or more whorls of bracts below and near a flower, flower cluster, or fruit (fig. 65.1)
keel: one part of a flower in the pea family; usually fused and encloses the stamens and pistil (fig. 65.3)
lanceolate: (leaf shape) formed like a lance: narrow, oval tapering to a point at each end (fig. 65.7)
lateral roots: roots that extend horizontally from the primary root
latex: milky fluid found in some plants; usually exuded after an injury to the plant as a protective measure; coagulates when exposed to air; natural rubber is an example of latex
leaf axil: acute angle between a shoot and a leaf stem or petiole extending from the shoot; buds develop here (fig. 65.10)
leaflet: leaflike part of a compound leaf; may resemble a complete leaf, but it is not borne on the main plant stem or branch (whereas a leaf is), rather it is on a petiole or a branch of the leaf (fig. 65.7)
leaf scar: mark on the twig after the leaf falls off
legumes: class of vegetables that includes beans, peas, and lentils; in the family Fabaceae
lenticels: multiple small pores/slits that allow gas exchange between the atmosphere and the internal tissues of a woody plant stem
lobes: (on leaf) rounded outgrowth that is part of a larger structure (fig. 65.4)
lyrate: (leaf shape) having or being in the shape of a lyre (fig. 65.7)
mericarps: part of a fruit that splits apart into several perfect fruit; can be seen in the schizocarp (dry fruit) of plants in the carrot family (fig. 65.12)
monoecious: plant where both the male and female reproductive organs are on the same individual; hermaphrodite
mycelium network: network of incredibly tiny threads/roots of a greater fungal organism. When these bore into tree roots, they make what is called a mycorrhizal network. These networks benefit both fungi and trees in the form of sharing of and access to water, nitrogen, carbon and other minerals. See figure 38.5 on page 142.
myco-heterotrophic: plants that partly or completely rely on fungi for their energy and nutrients
mycorrhiza: (mycorrhizas or mycorrhizae, plural) root system of a plant and the role that the fungus within it plays; important for the nutrition of plants and for the health of the soil
naturalized: plant having established in the wild in a region where it is not indigenous
nitrogen-fixing nodules: some plants (notably legumes) can form a symbiotic relationship with nitrogen-fixing soil bacteria called rhizobia within which the bacteria can convert atmospheric nitrogen into ammonia. When that plant withers and dies, the nitrogen is released and becomes accessible for other species via the soil and groundwater.
node: point on a stem where the buds, leaves, and

branching twigs grow from; between nodes are internodes (fig. 65.10)

nomenclature: system for classifying and naming things, especially in science or other discipline; rules for naming and classifying thing within that system

ochrea: dry sheath around a stem formed by the merging of two or more stipules; a characteristic we can observe in plants of the dock family [Polygonaceae]

opposite: leaf arrangement where two leaves develop on opposite sides of the stem by a node (fig. 65.10)

opposite decussate: where sets of leaf develop at a 90° angle from the leaf set below and above on stem (fig. 65.10)

ovary: female organ of a flower; enlarged structure located at the base of flower; within it are ovules, which develop into seeds upon fertilization; will mature into a fruit, either dry or fleshy, enclosing the seeds (fig. 65.1)

palmate: leaf with several lobes (typically 5 to 7), where the midribs all radiate from one point like fingers from a hand (fig. 65.7)

panicle: much-branched flower cluster (raceme) where each branch has more than one flower (fig 65.11)

pappus: tuft of hairlike or frilly appendages that help some seeds transport on the wind; the parachute-like hairy fluff attached to the dandelion seed (fig 65.12)

parasitic: living off another organism; a parasite is an organism that lives on or inside a host and obtains its food from the host to the host's detriment

parthenogenesis: type of reproduction in which a seed can be formed without the need for external pollen from another individual plant

pedicel: small stalk that attaches an individual flower to the main axis of the inflorescence (fig. 65.5)

peduncle: stalk that joins an entire inflorescence cluster to the main stalk (fig. 65.5)

perennial: plant that lives more than two years

peridium: outer layer that encloses a mass of spores in gasteroid fungi (fig. 65.8)

perithecia: (in some fungi) round or flask-shaped fruiting body with a pore where the spores disperse from (fig. 65.9)

petals: (often) brightly colored leaves surrounding the reproductive parts of a flower; attract pollinators (fig. 65.1)

petiole: slender stem that attaches the leaf to the stem or branch (fig. 65.5)

photosynthesis: process where the chlorophyll in a plant absorbs the sunlight. Plant uses the sun to change water (in the soil) and carbon dioxide (from the atmosphere) into oxygen and sugars. The plant uses some of the sugars and stores the rest. It releases oxygen back into the atmosphere.

phytoremediation: plant-based approach to extract and remove elemental pollutants in the soil

pinnately compound leaf: leaf that is divided into smaller leaflets arranged on each side of the leaf's central stalk (fig. 65.7); simple pinnate: leaflets are directly attached to the leaf's central stalk/main axis (the rachis) (fig. 65.7); double/bipinnate: twice divided, leaflets along secondary stem or rachis (fig. 65.7)

pioneer species: hardy plant species that are the first to colonize barren environments or areas disrupted by events such as wildfire

pistil: female reproductive part of a flower; usually made up of a swollen base (the ovary) and a hollow stalk, called a style, that starts at the ovary and ends in a pollen-receptive sticky tip (the stigma), usually at the center of the flower (fig. 65.1)

pitch: polymer that can be found naturally in plants or can be manufactured from petroleum or coal; also known as resin; can be turned into rosin

pith: also known as medulla; a tissue in the stems of vascular plants; made from soft spongy parenchyma cells; transports nutrients through the plant and stores nutrients within its cell

pollen: powdery substance produced by a male plant to fertilize a female plant; when it lands on a suitable pistil, a pollen tube is produced that transfers the pollen to the ovary where the seed is created

pome: type of fruit/berry that shares certain structural similarities; has an endocarp (tough fibrous flesh that surrounds the seeds), and a softer mesocarp (edible flesh) that encases the endocarp. The pome's epicarp (skin) protects the edible flesh. An example of a pome is an apple. (fig. 65.12)

prickles: short, skinny, sharp-pointed outgrowth from the bark or epidermis of a plant; a small thorn. Does not contain vascular tissue.

primocanes: slender stalks/canes in certain plants such as raspberries. First-year canes are called primocanes, which after a rest period (usually over winter in colder climates), flower and produce fruit the following year. See floricanes.

prostrate: (of a plant or stem) lying flattened to the ground

pseudo-stem: false stem made from the rolled bases of leaves

pubescent: (botany) term used to describe a plant that is covered by a layer of fine hair

raceme: inflorescence where each flower is borne on an individual stem branching from the main axis. The oldest flowers grow closest to the base, and new flowers are produced as the branch or stem grows in height. (fig. 65.11)

ray floret/ray petal: one of the narrow flowers that resemble single petals in plants from the Asteraceae family (sunflowers); surround the central disk of disk florets in the center of such flowers (fig. 65.3)

resin: certain plants release resin when wounded to protect themselves against pathogens and insect attack

reticulated: being marked like a net or network with interlacing lines

rhizome: also called creeping rootstalk; horizontal underground plant stems that can produce shoots and roots and develop a new plant (fig. 65.6)

rhomboid: (leaf shape) diamond shaped; has four sides (fig. 65.7)

ring (mushrooms): ringlike or collar-like structure that can be found on the stipe (stalk) of some types of mushrooms; also called annulus. It is left behind on the stipe after the partial veil that covers the entire young specimen ruptures to expose the spore-producing surfaces (such as the gills). (fig. 65.8).

riparian: strip of land between dry land and a body of water. Riparian plants are moisture loving and can be found near waterways. An important habitat to keep water healthy and clean.

rosette: see basal rosette (fig. 65.5)

rosin: also called colophony or Greek pitch; comes from the resin of pines and other conifers and is used in printing inks, photocopying, adhesives, and sealing

samara: type of dry winged fruit; maple seeds are samaras (fig. 65.12)

sap: fluid that transports water and nutrients through the xylem cells of a plant

sapling: young tree with a slender trunk

saprobic: feed on dead or decaying organic matter

sapwood: living part of a tree trunk where transportation of water and minerals take place; between the cambium and the heartwood; also called alburnum

scale (on cone): can be seen on coniferous cones; the seeds are held under the cone scales

secondary gills: see gills, short

seedlings: young plant. A typical young seedling has three main parts: the radicle (embryonic root), the hypocotyl (stem), and the cotyledons (baby leaves that emerge from the seed).

self-fertile: (of a plant) capable of self-fertilization; being able to make seeds without the help of pollen from a different plant

self-pruning: shedding of branches, limbs, or leaves as a response to an environmental situation such as disease, shade, or drought; done by a tree to protect itself and keep its resource output smaller

sepals: petals/leaves (often green) on the bottom of a flower and on the outside of a developing bud to protect it (fig. 65.1)

serotinous: a conifer cone that requires heat from a wildfire to open (in order to disperse its seed) is serotinous

serrated: (leaf edge) having a jagged edge; sawlike (fig 65. 4)

shoot: (botany) a plant shoot is a growing aboveground plant stem together with its leaves and lateral buds, flowering stems, and flower buds

spatulate: (leaf shape) broad at the apex and tapered at the base (fig. 65.7)

species: group of living organisms made up of similar individuals able to exchange genes or interbreed. The species is the principal natural taxonomic unit, ranking below a genus and denoted by a Latin binomial, e.g., *Homo sapiens* for humankind. The genus is *Homo* and the species is *Homo sapiens*. *Sapiens* shows that modern humans are a species within the genus *Homo*.

spike (flower): flower configuration where the flowers develop directly from the stem and lack pedicels (leaf stalks) (fig. 65.11)

spine: (botany) modified leaves, parts of leaves, or stipules that make a thornlike protrusion; contain vascular tissue

spores: (botany) unit of sexual or asexual reproduction. It is transferred from male plant/mushroom to female plant/mushroom in a variety of ways, such as by wind and insect pollinators. Spores are part of the life cycle of many plants, algae, fungi, and protozoa.

spore print: powdery image created by allowing spores of a mushroom to fall onto a surface. The spore print is used as one way to identify mushrooms since it can show both the color of the spores and, when viewed through a microscope, the shape of the spores.

spur: botanical term given to protrusions of tissue on certain plant organs; common in the *Viola* genus (fig. 65.3)

stalk: stem or main axis of a plant (fig. 65.5)

stamen: collective name for the male pollen-producing part of a flower; includes anther and filament (fig 65.1)

stand: collection of trees or other plants that occupy a specific area and are sufficiently uniform in species composition, size, age, arrangement, and condition to be distinguished from the forest or other plants growing around them

stigma: part of a plant's female reproductive organ. The stigma receives pollen during pollination, and the pollen grains germinate here. The stigma is adapted to catch its species' pollen through sticky surfaces, hair, and shape. (fig. 65.1)

stipitate glands: glands on the ends of stalks

stipules: small leaflike appendages to a leaf, typically seen in pairs at the base of the leaf stalk (fig. 65.5)

stomata: (botany) (stoma, singular; stomata, plural) pores in the epidermis (outer layer) of leaves, stems, and other organs that control the rate of gas exchange

stomatal band: lighter band on needles responsible for the intake of carbon dioxide and release of oxygen in some conifer species

strobili: (strobilus, singular) plant structure that holds spores along a central axis (stem) with spirally arranged leaves or modified stems surrounding it; can be seen in the genus *Equisetum*

style: slender stalk connecting the ovary and stigma; part of a plant's female reproductive parts (fig. 65.1)

subgleba: below the gleba, the main body that produces the spores in some basidiomycetes (fig 65.8)

sucker: vegetative stem structure. Suckers start from underground stems and grow toward the surface.

supporting roots: roots that extend laterally to give a tree or bush more stability

symbiotic relationship: term describing relationship or interaction between two differing organisms. It can be either mutualistic (both parties receive benefit), commensalistic (one party receives benefit without loss to the other), or parasitic (one benefits to the detriment of the other).

taproot: main, large and central root from which other roots sprout sideways (fig. 65.6)

tar: wood tar is a black sticky substance extracted from conifer trees as well as deciduous trees like birch and beech. It has been used in soaps, shampoos, and medicine and for waterproofing and hoof care.

taxonomy: branch of science that deal with classification; classification of organisms

tepal: term used when the sepals and petals are hard to classify as either sepals or petals; collective name for both

terminal bud: bud located at the tip of the stem and where most of the plant growth occurs

thorn: spiky modified branch or stem; may be simple or branched and has vascular bundles inside

trifoliate: (leaf shape) having three leaflets (fig. 65.7)

tuberculate ectomycorrhiza: structure where ectomycorrhiza are enclosed within a fungal hyphae; ectomycorrhiza is a symbiotic relationship between a fungus and a plant and give each added benefits such as greater access to nutrition, water, and protection from disease

umbel: flower cluster where stalks of nearly equal length start from a common center and form a flat or curved surface; common in the carrot family [Umbelliferae] (fig. 65.11)

umbo: raised area in the center of a mushroom cap (fig. 65.8)

understory: layer of vegetation beneath the main canopy of a forest, consisting of shrubbery, herbs, and smaller trees

veil: also called velum; one of several structures in fungi, here we refer to the thin membrane that covers the cap and stalk of an immature mushroom (fig. 65.8)

venation: the arrangement of veins in a leaf

volva: (mycology) a cuplike structure at the base of a mushroom (fig. 65.8)

watershed: the area of land that funnels water into a specific waterbody, e.g., streams, lakes, oceans, or bays; every body of water has a watershed surrounding it

whorl: (botany) an arrangement of leaves, sepals, petals, stamens, or carpels that fan out from a single point and surround or wrap around the stem or stalk; a whorl needs to have at least three leaves to be called a whorl; a pair of leaves does not constitute a whorl (fig. 65.10)

wind pollination: also called anemophily; the process when pollen from male plant is transferred to female plant (to reproduce) via air currents

Glossary Drawings

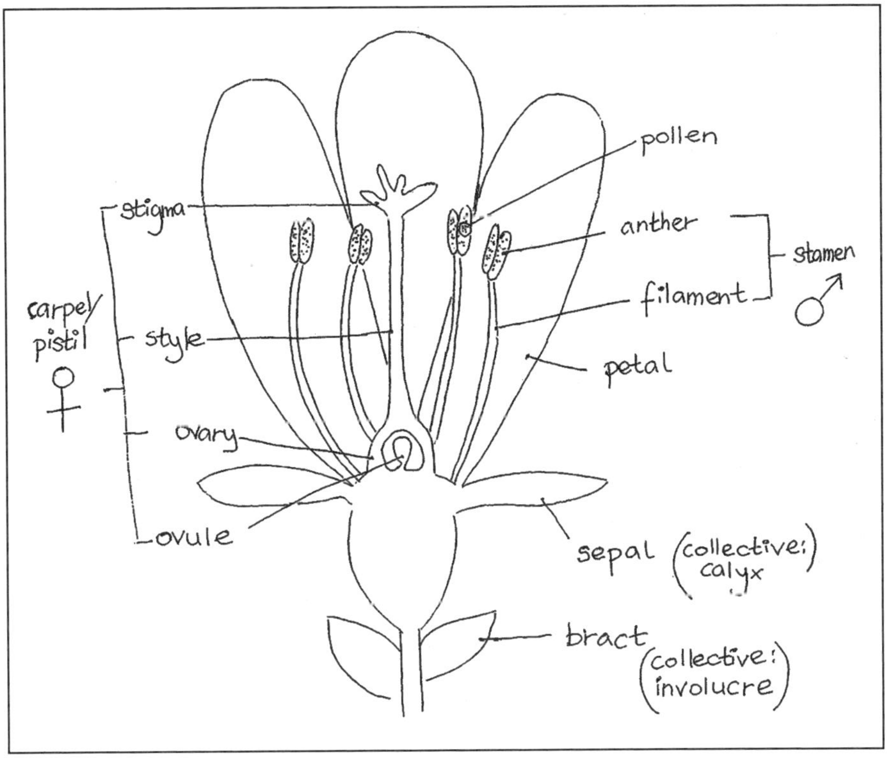

Figure 65.1. Flower reproductive parts.

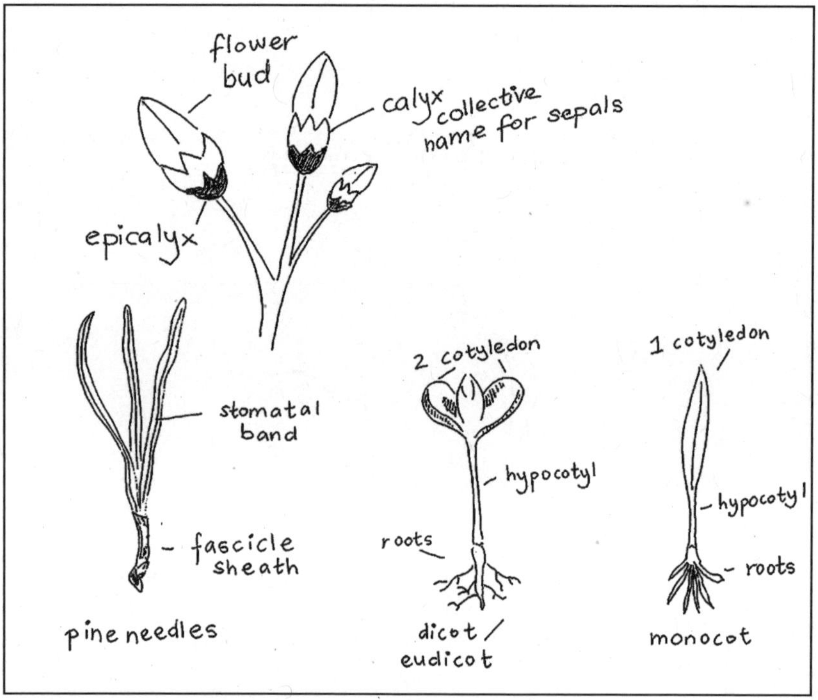

Figure 65.2. Bud, needle, cotyledon.

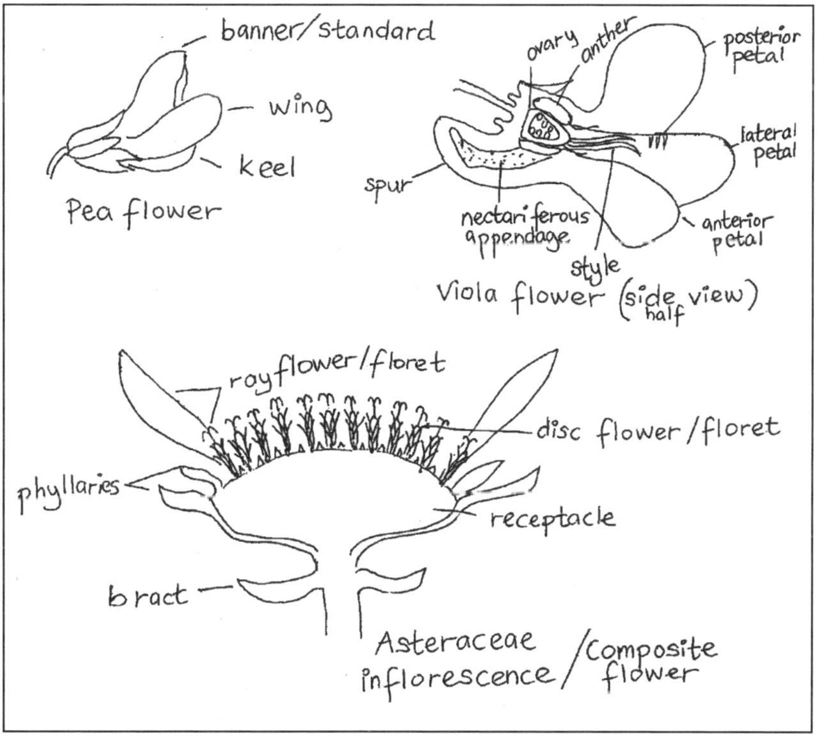

Figure 65.3. Top left to bottom: pea flower diagram, viola flower diagram, and aster flower diagram.

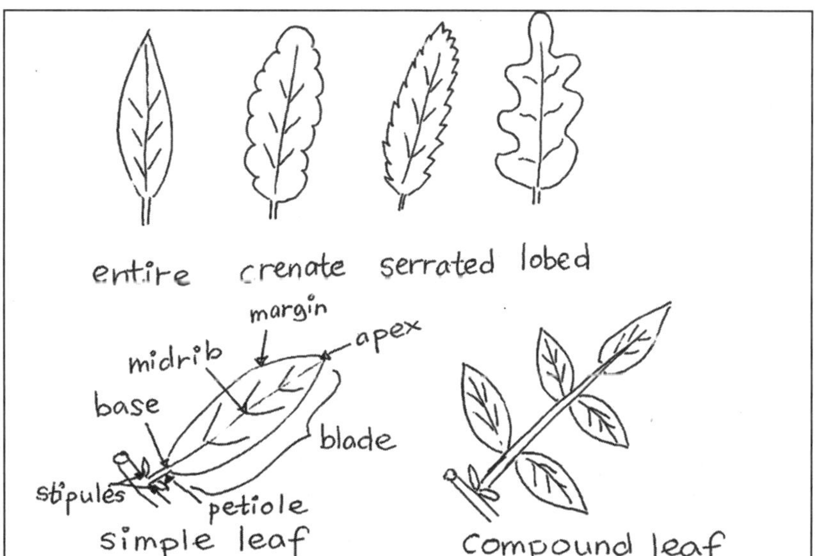

Figure 65.4. Leaf edges and names of leaf parts.

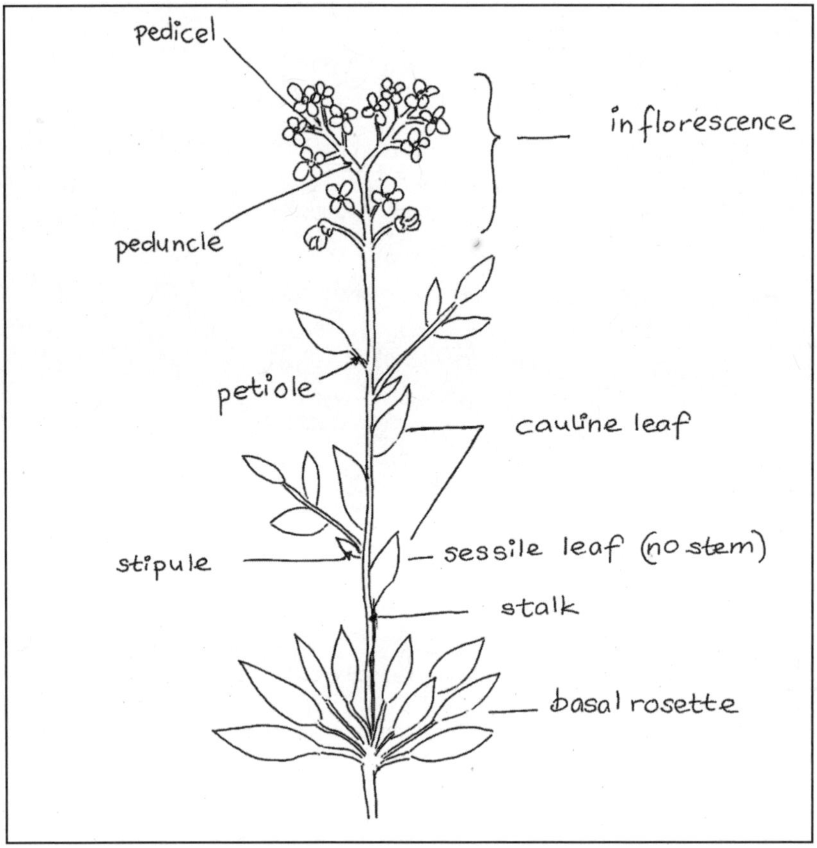

Figure 65.5. Parts of plant.

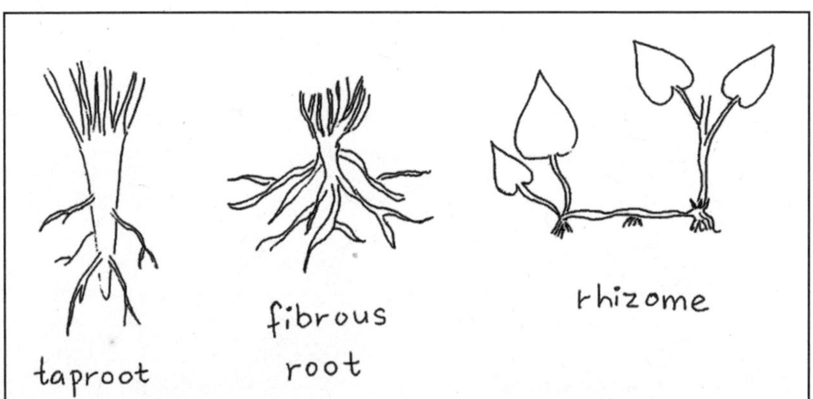

Figure 65.6. Types of roots.

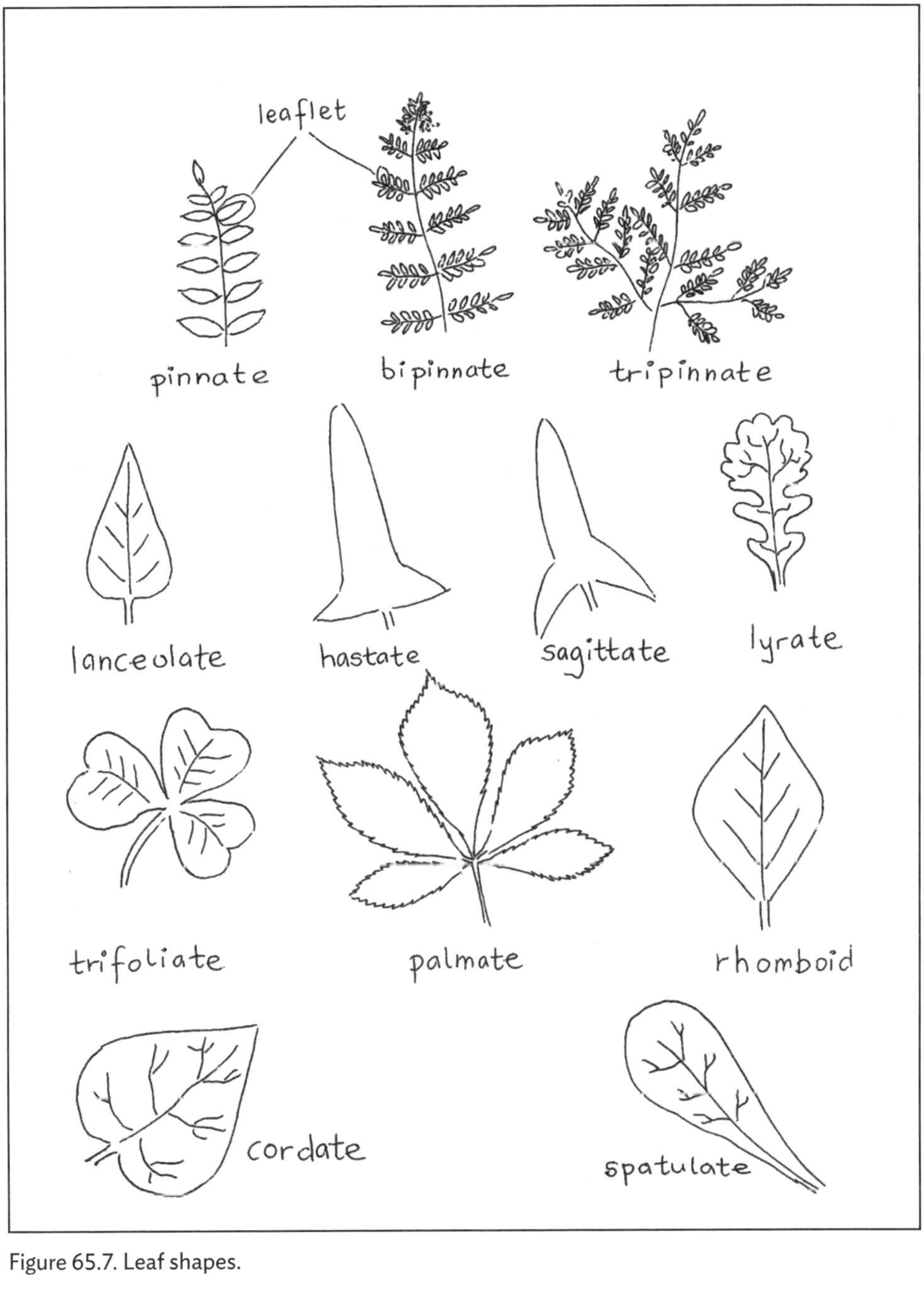

Figure 65.7. Leaf shapes.

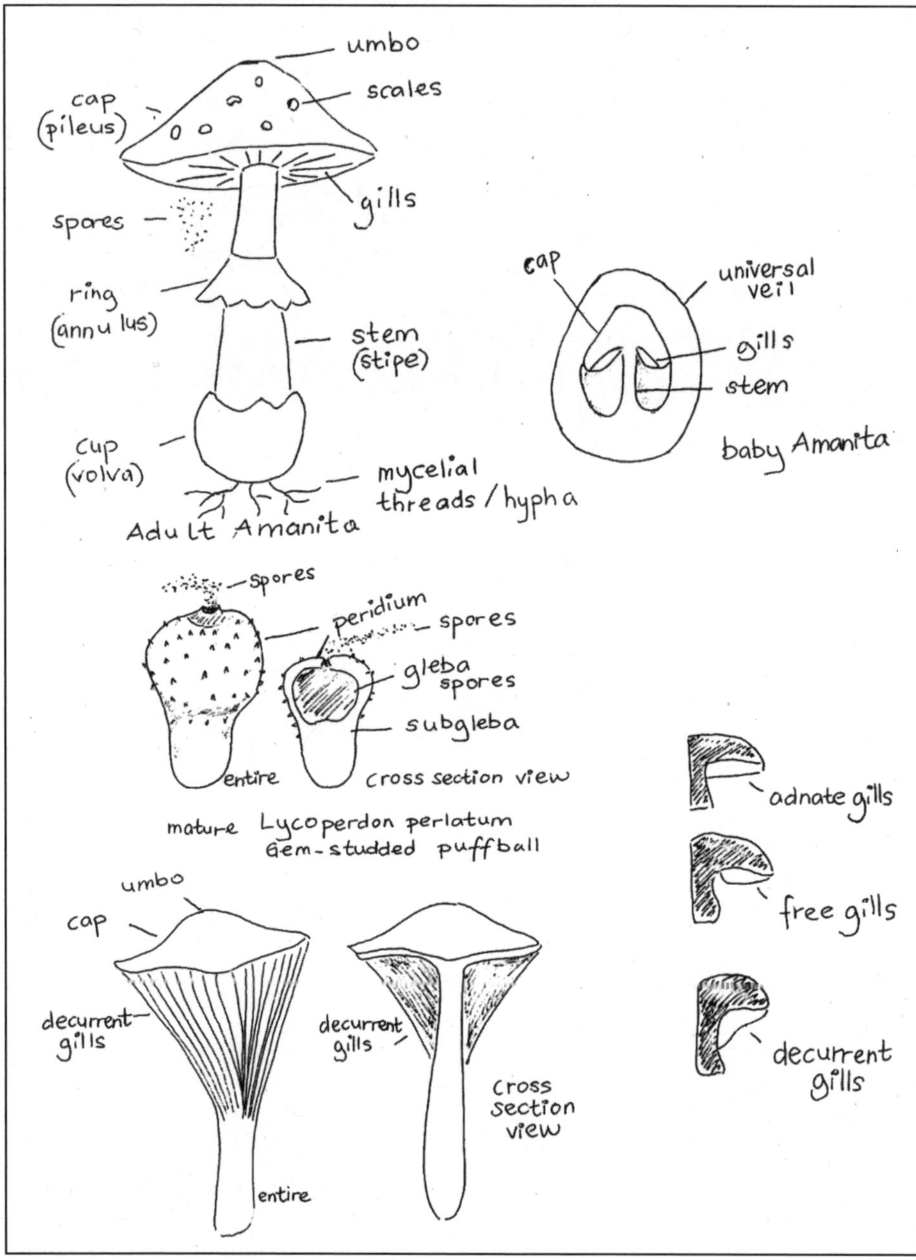

Figure 65.8. Mushroom parts.

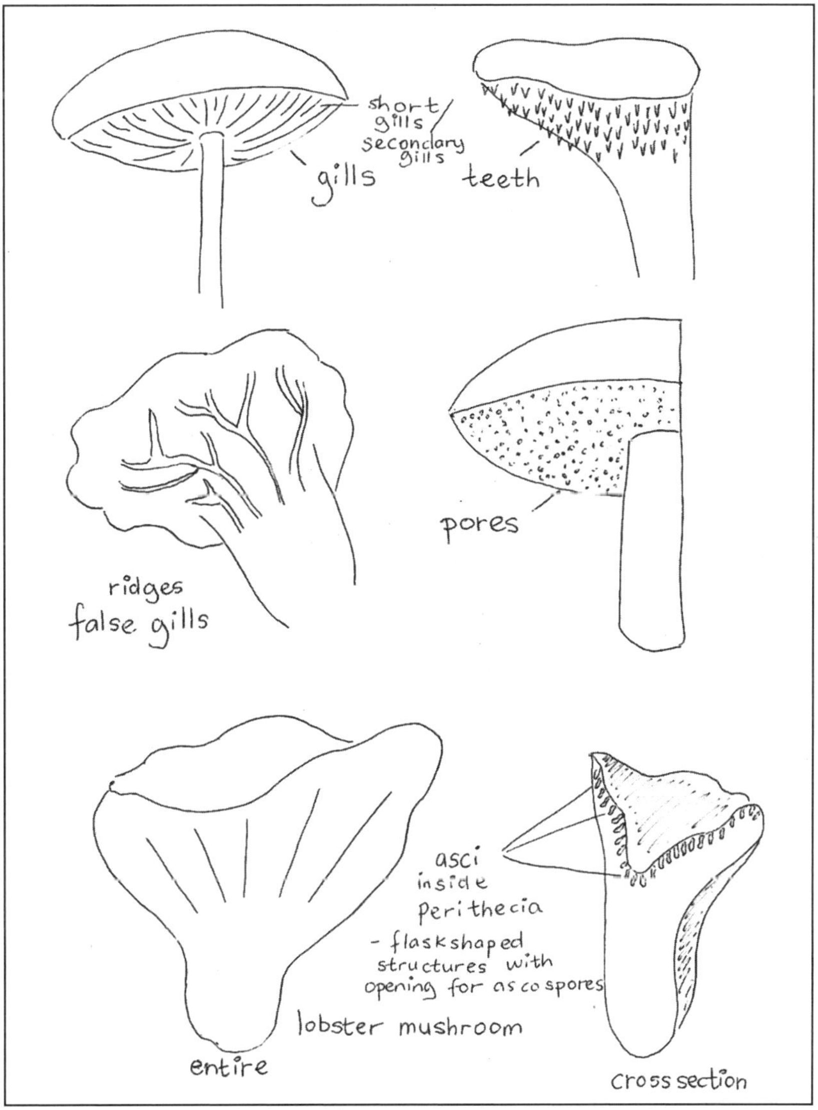

Figure 65.9. Gills and other spore producing organs.

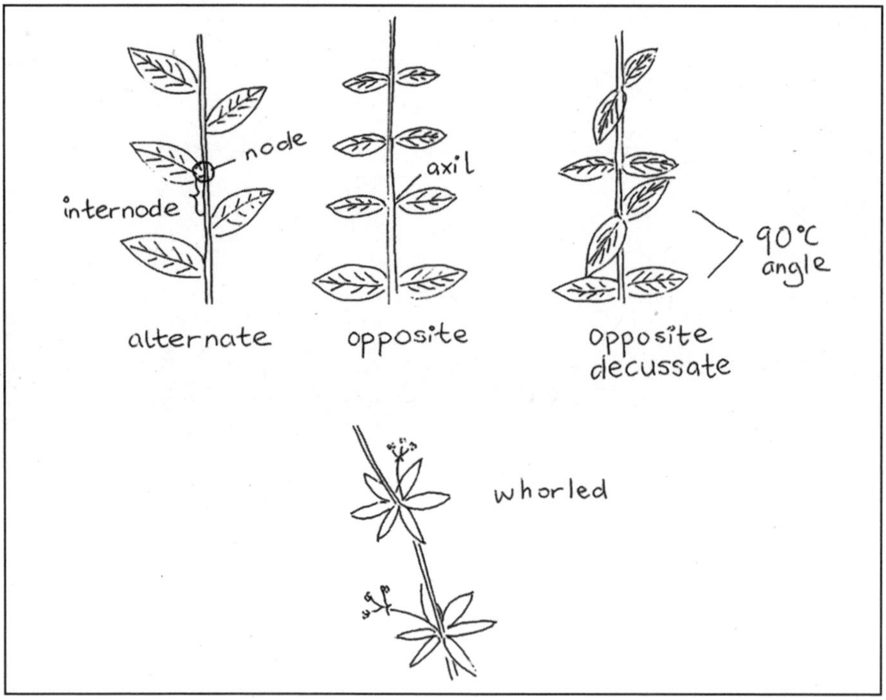

Figure 65.10. Leaf arrangement.

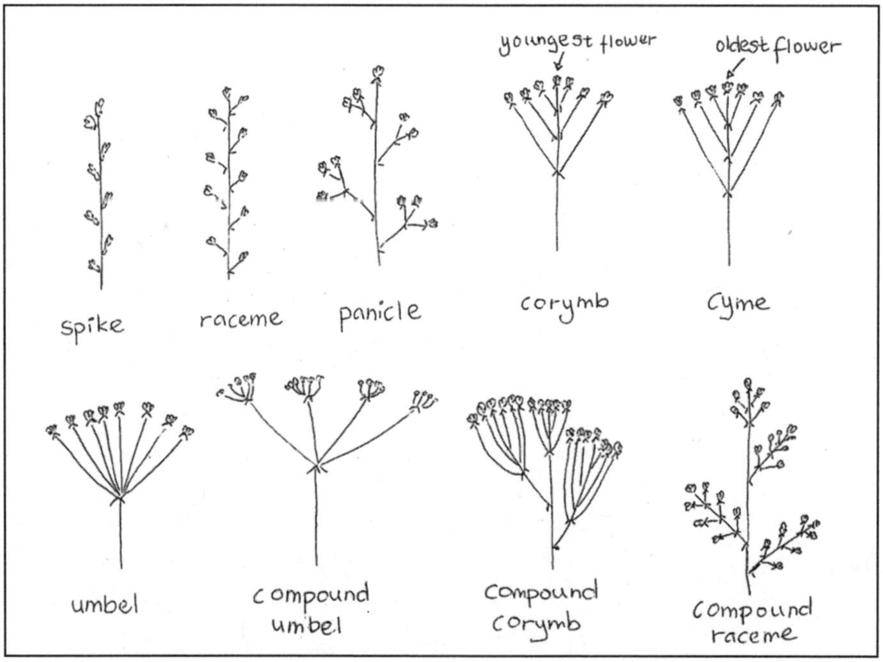

Figure 65.11. Flower arrangement.

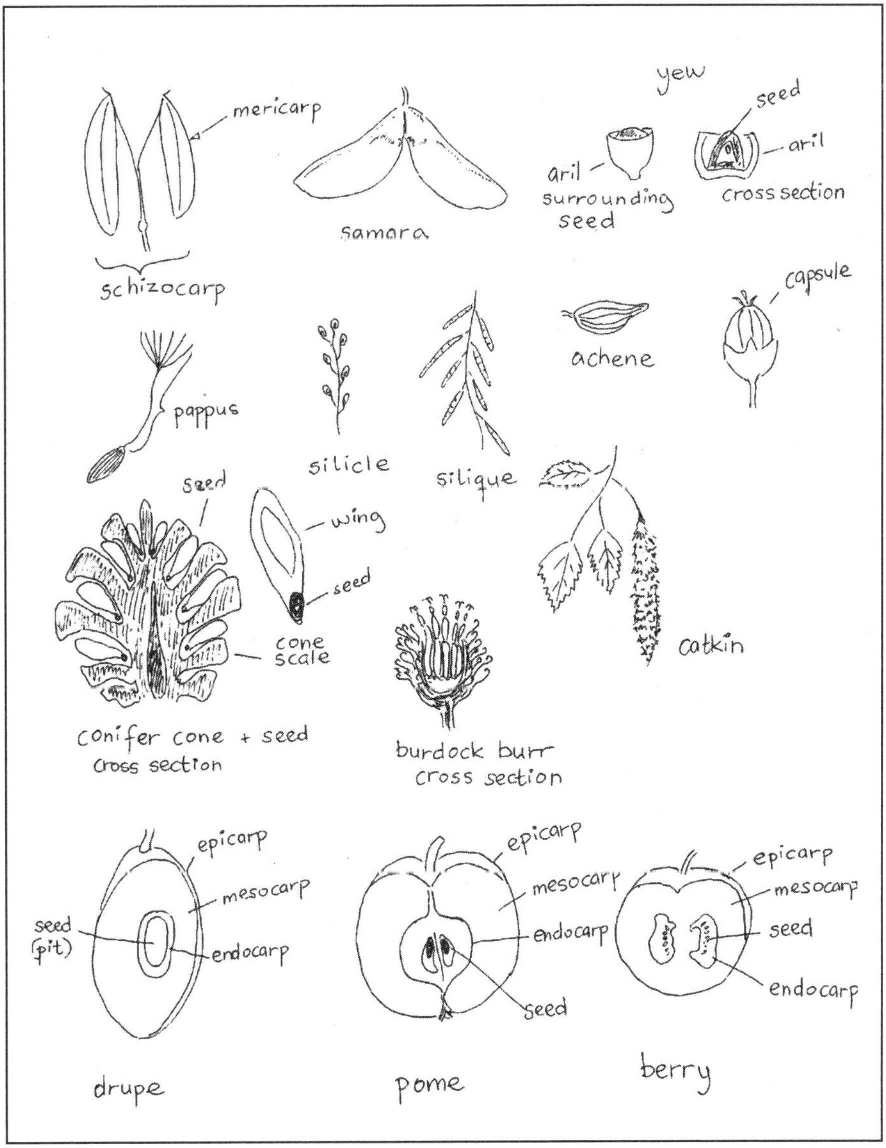

Figure 65.12. Fruits and seeds.

References

Applequist, W.L., and D.E. Moerman. 2011. "Yarrow (*Achillea millefolium* L.): A Neglected Panacea? A Review of Ethnobotany, Bioactivity, and Biomedical Research," *Economic Botany*, 65, 209.

Balch, Phyllis A. 2002. *Prescription for Herbal Healing*. New York: Avery, Penguin Putnam.

Brayshaw, Christopher. 1996. *Catkin-Bearing Plants of British Columbia*. Victoria, BC: Royal British Columbia Museum.

Chidrawar, V.R., K.N. Patel, N.R. Sheth, S.S. Shiromwar, and P. Trivedi. 2011. "Antiobesity Effect of Stellaria media Against Drug Induced Obesity in Swiss Albino Mice," *Ayu*, 32(4): 576–584.

Chisholm, Libby Jay. 2012. "Eating Our Culture: Intersections of Culturally Grounded Values-Based Frameworks and Indigenous Food Systems Restoration in Secwepemcúlecw." BA thesis, University of Guelph, ON.

Department of Northern Affairs and National Resources. 1956. *Native Trees of Canada*. Ottawa: Forestry Branch.

Domico, Terry. 2008. *Wild Harvest: Edible Plants of the Pacific Northwest*. Surrey, BC: Hancock House.

Duke, James A. 1997. *The Green Pharmacy*. New York: St Martin's Paperbacks.

Elpel, Thomas. J. 2013. *Botany in a Day: The Patterns Method of Plant Identification*. Pony, MT: Hops Press.

Festschrift, Prithipalsingh. 2012. *Plant Taxonomy: Past, Present, and Future*. Delhi: Energy and Resources Institute.

Foster, Steve, and Christopher Hobbs. 2002. *Western Medicinal Plants and Herbs*. Boston: Houghton Mifflin.

Gray, Beverly. 2011. *The Boreal Herbal: Wild Food and Medicine Plants of the North*. Whitehourse: Aroma Borealis Press.

Green, James. 2000. *The Herbal Medicine-Maker's Handbook: A Home Manual*. Toronto: Crossing Press.

Golmohammadi, Farhood, and Mohammad Karim Motamed. 2012. "A Viewpoint Toward Farm Management and Importance of Barberry in Sustainable Rural Livelihood in Desert Regions in East of Iran," *African Journal of Plant Science*, 6(7): 213–221.

Hoffman, David. 1990. *The New Holistic Herbal*. Shaftesbury, UK: Element.

Ignace, Marianne, and Ronald E. Ignace. 2017. *Secwépemc People, Land and Laws. Yeri7 re Stsqeys-kucw*. Montreal: McGill-Queen's University Press.

Joseph, Bob. 2018. *21 Things You May Not Know About the Indian Act: Helping Canadians Make Reconciliation with Indigenous Peoples a Reality*. Indigenous Relations Press.

Keane, Kahlee. 2015. *The Standing People: Wild Medicinal Plants of British Columbia*. British Columbia: Save Our Species.

Kim, J.H., et al. 2012. "Immunomodulating and Antitumor Activities of *Panellus serotinus* Polysaccharides," *Mycobiology*, 40(3): 181–188.

Kimmerer, Robin Wall. 2013. *Braiding Sweetgrass. Indigenous Wisdom, Scientific Knowledge and the Teachings of Plants*. Minneapolis: Milkweed Editions.

Kloos, Scott. 2017. *Pacific Northwest Medicinal Plants: Identify, Harvest and Use 120 Wild Herbs for Health and Wellness*. Portland, OR: Timber Press.

Lim, G. 2017. "Warfarin: From Rat Poison to Clinical Use," *Nature Reviews Cardiology*.

Lincoff, Gary H. 1981. *The Audubon Society Field Guide to North American Mushrooms*. New York: Alfred A. Knopf.

Lincoff, Gary. 2017. *The Complete Mushroom Hunter: An Illustrated Guide to Foraging, Harvesting and Enjoying Wild Mushrooms*. Minneapolis: Quarto Publishing.

Lara, Maria V., Maria F. Drincovich, and Carlos S. Andreo. 2004. "Induction of a Crassulacean Acid-like Metabolism in the C4 Succulent Plant, *Portulaca oleracea* L.: Study of Enzymes Involved in Carbon Fixation and Carbohydrate Metabolism," *Plant and Cell Physiology*, 45(5): 618–626.

Lawless, Julia. 2014. *The Encyclopedia of Essential Oils: The Complete Guide to the Use of Aromatic Oils in Aromatheraphy, Herbalism, Health & Well-being*. London, UK: Harper Thorsons.

MacKinnon, A. L. Kershaw, J. T. Arnason, P. Owen, A. Karst, and F. Hamersley-Chambers. 2009. *Edible & Medicinal Plants of Canada*. Edmonton, AB: Lone Pine.

Mase, Guido. 2013. *The Wild Medicine Solution: Healing with Aromatic, Bitter and Tonic Plants*. Toronto: Healing Arts.

McKenny, Margaret, and Daniel E. Stuntz. 1987. *The New Savory Wild Mushroom*. Vancouver, BC: Greyston.

Meredith, Leda. 2019. *The Skillful Forager: Essential Techniques for Responsible Foraging and Making the Most of Your Edibles*. Boulder, CO: Roost Books.

Mohamed, A.I., and A.S. Hussein. 1994. "Chemical Composition of Purslane (*Portulaca oleracea*)," *Plant Foods for Human Nutrition*, 45: 1–9.

Neal, Bill. 1992. *Gardener's Latin*. Chapel Hill, NC: Algonquin Books.

Neddo, Nick. 2015. *The Organic Artist: Make Your Own Paint, Paper, Pigments, Prints and More from Nature*. Beverly, MA: Quarry Books.

Oladeji, O. S., and A. K. Oyebamiji. 2020. "*Stellaria media* (L.) Vill.—A Plant with Immense Therapeutic Potentials: Phytochemistry and Pharmacology," *Heliyon*. 6(6).

Orlando, Richard. 2018. *Weeds in the Urban Landscape: Why They Are Here and How to Live with Them*. Berkeley, CA: North Atlantic Books.

Parish, Roberta. 1960. *Tree Book: Learning to Recognize Trees of British Columbia*. Victoria, BC: Canadian Forest Service.

Parish, Roberta, Ray Coupe, and Dennis Lloyd, (Eds.). 1996. *Plants of Southern Interior British Columbia*. Vancouver, BC: Lone Pine.

Poltavets, Svetlana, and Eugene Poltavets. 2020. *A Guide to Medicinal Mushrooms of the Pacific Northwest: Health Benefits and Other Therapeutic Uses*. Surrey, BC: Hancock House.

Prank, Marje, Shawn C. Kenaley, Gary C. Bergstrom, Maricelis Acevedo, and Natalie M. Mahowald. 2019. "Climate Change Impacts the Spread of Wheat Stem Rust, a Significant Crop Disease," *Environmental Research Letters*, 14(12).

Raygorodetsky, Gleb. "Indigenous Peoples Defend Earth's Biodiversity—but They're in Danger," *National Geographic*, November 16, 2018.

Richter, Dana L., and Bradley L. Morse. 2008. "The Western Giant Puffball (*Calvatia booniana*) in Northern Michigan," *The Great Lakes Botanist: A Journal of North American Botany*, 47(2).

Rogers, Robert. 2011. *The Fungal Pharmacy: The Complete Guide to Medicinal Mushrooms and Lichens of North America*. Berkeley, CA: North Atlantic Books.

Romm, Aviva, Mary L. Hardy, and Simon Mills. 2010. *Botanical Medicine for Women's Health*. London, UK: Churchill, Livingstone, Elsevier.

Royer, France, and Richard Dickinson. 2006. *Weeds of Canada and the Northern United States*. Edmonton, AB: Lone Pine.

Saunders, Diane G.O., Zacharias A. Pretorius, and Mogens S. Hovmoller. 2019. "Tackling the Re-emergence of Wheat Stem Rust in Western Europe," *Communications Biology*, 2: 51.

Schalkwijk-Barendsen, Helene. 1991. *Mushrooms of Western Canada*. Edmonton, AB: Lone Pine.

Schuster, Richard, Ryan R. Germain, Joseph R. Bennett, Nicholas J. Reo, and Peter Arcese. 2019. "Vertebrate Biodiversity on Indigenous-managed Lands in Australia, Brazil, and Canada Equals That in Protected Areas," *Environmental Science & Policy*, 101: 1 6.

Scott, Julian. 1990. *Natural Medicine for Children*. London, UK: Gaia Books.

Sept, J. Duane. 2006. *Common Mushrooms of the Northwest*. Sechelt, BC: Calypso.

Zhao, Shiyang, and Erbilgin Nadir. 2019. "Larger Resin Ducts Are Linked to the Survival of Lodgepole Pine Trees During Mountain Pine Beetle Outbreak," *Frontiers in Plant Science*, 10: 1459.

Simard, Suzanne. 2021. *Finding the Mother Tree: Discovering the Wisdom of the Forest*. New York: Allen Lane, Penguin Random House.

Thayer, Samuel. 2017. *Incredible Wild Edibles: 36 Plants That Can Change Your Life*. Bruce, WI: Forager's Harvest Press.

Tis Mal Crow. 2000. *Native Plants, Native Healing: Traditional Mukogee Way*. Summertown, TN: Native Voices.

Treu, Roland, et al. 2014. "Decline of Ectomycorrhizal Fungi Following a Mountain Pine Beetle Epidemic," *Ecology*, 95(4): 1096–1103.

Turner, Nancy J. 1979. *Plants in British Columbia Indian Technology*. Victoria, BC: Royal British Columbia Museum.

Turner, Nancy J. 2014. *Ancient Pathways, Ancestral Knowledge. Ethnobotany and Ecological Wisdom of Indigenous Peoples of Northwestern North America*. Vol 1 & 2. Montreal: McGill-Queen's University Press.

Turner, Nancy J., Laurence C. Thompson, M. Terry Thompson, and Anne Z. York. 1990. *Thompson Ethnobotany: Knowledge and Usage of Plants by the Thompson Indians of British Columbia*. Victoria, BC: Royal British Columbia Museum.

Turner, Nancy J., R. Bouchard, and Dorothy I. D. Kennedy. 1980. *Ethnobotany of the Okanagan-Colville Indians of British Columbia and Washington*. Victoria, BC: Royal British Columbia Museum.

UBC News. 2019. "Biodiversity Highest on Indigenous-managed Land. UBC-led Study Highlights Importance of Collaborating with Indigenous Communities to Protect Species." https://news.ubc.ca/2019/07/31/biodiversityhighest-on-indigenous-managed-lands

Waxman, Olivia B. "The History of Native American Boarding Schools Is Even More Complicated Than a New Report Reveals. *Time Magazine*, May 17, 2022.

Wiley, Erin, Bruce J. Rogers, Robert Hodgkinson, and Simon M. Landhausser. 2015. "Nonstructural Carbohydrate Dynamics of Lodgepole Pine Dying from Mountain Pine Beetle Attack," *New Phytologist Foundation*, 209(2): 550–562.

Yost, John. 2016. *How to Make Birch Bark Baskets: Wilderness Survival Skills Series*. Charleston, SC: CreateSpace.

Young, Devon. 2019. *The Backyard Herbal Apothecary: Effective Medicinal Remedies Using Commonly Found Herbs and Plants*. Salem, MA: Paige Street.

Zachos, Ellen. 2013. *Backyard Foraging: 65 Familiar Plants You Didn't Know You Could Eat*. North Adams, MA: Storey Publishing.

Websites

American Conifer Society. Pinus ponderosa/ Ponderosa pine. https://conifersociety.org /conifers/pinus-ponderosa/

American Association of Equine Practitioners. Dr. Tom Lenz. https://aaep.org/horsehealth/red -maple-poisoning

Backyard Forager. Ellen Zachos. https://backyard forager.com

BC Species and Ecosystems Explorer. BC Government. Species-at-Risk Search Tool. https://a100 .gov.bc.ca/pub/eswp/search.do?method=change

Botanical.com. *A Modern Herbal* by Mrs. Maud Grieve. www.botanical.com

Conifers of the Palomar College Arboretum. www2 .palomar.edu/users/warmstrong/arbimg4a.htm

Convention on Biological Diversity. About Climate Change and Biological Diversity. www.cbd.int /climate/intro.shtml

Dave's Garden Botanary. https://davesgarden.com /guides/botanary/

Dr. Ryan Drum 2020. http://ryandrum.com/articles .htm

Drugs.com. Know more, be sure. www.drugs.com /cdi/bearberry.html

Eat the Weeds and Other Things, Too. Green Deane. www.eattheweeds.com

First Voices: www.firstvoices.com

Healing Mushrooms. https://healing-mushrooms.net /Lobster-Mushroom

Honest Food Net: Hunter, Angler, Gardener, Cook. https://honest-food.net/

Identify That Plant. Confidently master the skill of correct plant identification. Angelyn Whitmeyer. http://identifythatplant.com/three-easily-mixed -up-early-spring-plants

Illinois Wild Flowers. http://illinoiswildflowers.info/

Ministry of Environment, Okanagan Region. Government of British Columbia. Habitat Atlas for Wildlife at Risk. About Species at Risk. www.env .gov.bc.ca/okanagan/esd/atlas/atrisk.html

Minnesota Wildflowers: A Field Guide to the Flora of Minnesota. www.minnesotawildflowers.info/

Merck Manual: Veterinary Manual. www.merckvet manual.com

Mushroom Expert.com. www.mushroomexpert.com /gomphidius.html

Napiers. Monica Wilde. https://napiers.net/blogs /news/nettle-seed-uses

National Centre for Truth and Reconciliation: https://nctr.ca

Native American Ethnobotany. A Database of Foods, Drugs, Dyes and Fibers of Native American Peoples, Derived from Plants. http://naeb .brit.org

Native Land Digital: https://native-land.ca

Natural Resources Canada. Government of Canada. Trees, fact sheet. https://tidcf.nrcan.gc.ca/en /trees/factsheet/389

Naturally Wood. Trembling Aspen. www.naturally wood.com/species/trembling-aspen/

Newland, Bryan. 2022. *Federal Indian Boarding School Initiative Investigative Report.* www. bia.gov/sites/default/files/dup/inline-files/bsi _investigative_report_may_2022_508.pdf

Online Etymology Dictionary. www.etymonline.com /word/betula

Plants for a Future. https://pfaf.org/user/Default.aspx

Pleasant Valley Wetland Heritage Park. BC Small Wetlands Association. www.smallwetlands.com

Practical Self Reliance: Your Practical Guide to Self Reliant Living. https://practicalselfreliance.com/category/foraging

Pull Up Your Plants: The Ethnobotany of Foraged Food and Peculiar Produce. www.pullupyourplants.com/

Richters. Wild Ginger (*Asarum canadense*) and Aristolochic Acid. www.richters.com/show.cgi?page=QandA/Medicinal/20020313-4.html

RX list. Hemlock spruce interaction with lithium medication. www.rxlist.com/eastern_hemlock/supplements.htm

Sciencing. https://sciencing.com

Secwepemc-Kuc. Let the Journey Begin. http://secwepemc.sd73.bc.ca

Splatsin Tsm7aksaltn Splatsin Teaching Centre Society. www.splatsin.org

Tk'emlúps te Secwepémc. https://tkemlups.ca/profile/history/our-land

Tom Volk's Fungus of the Month. https://botit.botany.wisc.edu/toms_fungi/aug2001.html

Top Crop Manager. Carolyn King. 2015. The Barberry Connection with Stem Rust. www.topcropmanager.com/thebarberry-connection-with-stem-rust-17000

USDA Fire Effects Information System: Index of Species Information. www.fs.fed.us/database/feis/plants/tree/piceng/all.html

USDA Food Data Central Search Results. Nutrition fact sheets. US Department of Agriculture, Agriculture Research Centre. https://fdc.nal.usda.gov/fdc-app.html#/food-details/1102653/nutrients

USDA Natural Resources Conservation Service. Plant Guide. https://plants.usda.gov/java

USDA Food Nutritional Analysis, Agricultural Research Service. National Nutrient Database for Standard Reference Legacy Release. https://ndb.nal.usda.gov/ndb/foods/show/2960

Utah State University. Range Plants of Utah. https://extension.usu.edu/rangeplants/shrubs-and-trees/MapleRockyMountain

Utah State University Extension. https://extension.usu.edu/rangeplants/shrubs-and-trees/cedar_red

Western Native Seed. www.westernnativeseed.com

Wild Foods and Medicine. Elise Krohn. http://wildfoodsandmedicines.com/category/blog

Wild Food Girl: Foraging the wild for plants and stuff to eat. Erica M. Davis. https://wildfoodgirl.com

Index to Common and Scientific Names

Primary entries are bolded

About the Author

Mikaela is obsessed with plants and has travelled the world learning about them. Now, settled in the Okanagan, she loves teaching people what she knows about plants and their uses. When she is not teaching, you might find her studying new plants, concocting new wild recipes, enjoying time with her family, or being outdoors hiking, biking, skiing, or sailing. She is passionately engaged with her community and involved with various groups that work toward Truth and Reconciliation, environmental protection, and climate action.

To find out more about classes, newsletters, and events by Mikaela visit:
Website: www.forestfables.ca
Blog: foragingforestfables.blogspot.com
Instagram: @foragingasawayoflife

Credit: Peter Glimm

ABOUT NEW SOCIETY PUBLISHERS

New Society Publishers is an activist, solutions-oriented publisher focused on publishing books to build a more just and sustainable future. Our books offer tips, tools, and insights from leading experts in a wide range of areas.

We're proud to hold to the highest environmental and social standards of any publisher in North America. When you buy New Society books, you are part of the solution!

At New Society Publishers, we care deeply about *what* we publish—but also about *how* we do business.

- This book is printed on 100% **post-consumer recycled paper**, processed chlorine-free, with low-VOC vegetable-based inks (since 2002).
- Our corporate structure is an innovative employee shareholder agreement, so we're one-third employee-owned (since 2015)
- We've created a Statement of Ethics (2021). The intent of this Statement is to act as a framework to guide our actions and facilitate feedback for continuous improvement of our work
- We're carbon-neutral (since 2006)
- We're certified as a B Corporation (since 2016)
- We're Signatories to the UN's Sustainable Development Goals (SDG) Publishers Compact (2020–2030, the Decade of Action)

To download our full catalog, sign up for our quarterly newsletter, and to learn more about New Society Publishers, please visit newsociety.com

ENVIRONMENTAL BENEFITS STATEMENT

New Society Publishers saved the following resources by printing the pages of this book on chlorine free paper made with 100% post-consumer waste.

TREES	WATER	ENERGY	SOLID WASTE	GREENHOUSE GASES
67 FULLY GROWN	**5,400** GALLONS	**28** MILLION BTUs	**230** POUNDS	**28,900** POUNDS

Environmental impact estimates were made using the Environmental Paper Network Paper Calculator 4.0. For more information visit www.papercalculator.org

Certified B Corporation

new society PUBLISHERS www.newsociety.com

FSC www.fsc.org MIX Paper | Supporting responsible forestry FSC® C016245

SDG PUBLISHERS COMPACT